Publicly Funded Agricultural Research and the Changing Structure of U.S. Agriculture

Committee to Review the Role of Publicly Funded Agricultural Research on the Structure of U.S. Agriculture

BOARD ON AGRICULTURE AND NATURAL RESOURCES
DIVISION ON EARTH AND LIFE STUDIES
NATIONAL RESEARCH COUNCIL

NATIONAL ACADEMY PRESS
Washington, D.C.

NATIONAL ACADEMY PRESS   2101 Constitution Avenue, N.W.   Washington, D.C. 20418

NOTICE: The project that is the subject of this report was approved by the Governing Board of the National Research Council, whose members are drawn from the councils of the National Academy of Sciences, the National Academy of Engineering, and the Institute of Medicine. The members of the committee responsible for the report were chosen for their special competences and with regard for appropriate balance.

This study was supported by Contract/Grant No. 43-3AEL-7-80055 between the National Academy of Sciences and the U.S. Department of Agriculture Economic Research Service. Any opinions, findings, conclusions, or recommendations expressed in this publication are those of the author(s) and do not necessarily reflect the views of the organizations or agencies that provided support for the project.

International Standard Book Number 0-309-07616-1

Additional copies of this report are available from National Academy Press, 2101 Constitution Avenue, N.W., Lockbox 285, Washington, D.C. 20055; (800) 624-6242 or (202) 334-3313 (in the Washington metropolitan area); Internet, http://www.nap.edu

Suggested citation: National Research Council, 2001. *Publicly Funded Agricultural Research and the Changing Structure of U.S. Agriculture.* Committee to Review the Role of Publicly Funded Agricultural Research on the Structure of U.S. Agriculture (Washington, DC: National Academy Press).

Cover illustration reprinted, with permission, from Good Directions, Inc. Copyright 1996 by Good Directions, Inc.

Copyright 2002 by the National Academy of Sciences. All rights reserved.

Printed in the United States of America

THE NATIONAL ACADEMIES

National Academy of Sciences
National Academy of Engineering
Institute of Medicine
National Research Council

The **National Academy of Sciences** is a private, nonprofit, self-perpetuating society of distinguished scholars engaged in scientific and engineering research, dedicated to the furtherance of science and technology and to their use for the general welfare. Upon the authority of the charter granted to it by the Congress in 1863, the Academy has a mandate that requires it to advise the federal government on scientific and technical matters. Dr. Bruce M. Alberts is president of the National Academy of Sciences.

The **National Academy of Engineering** was established in 1964, under the charter of the National Academy of Sciences, as a parallel organization of outstanding engineers. It is autonomous in its administration and in the selection of its members, sharing with the National Academy of Sciences the responsibility for advising the federal government. The National Academy of Engineering also sponsors engineering programs aimed at meeting national needs, encourages education and research, and recognizes the superior achievements of engineers. Dr. Wm. A. Wulf is president of the National Academy of Engineering.

The **Institute of Medicine** was established in 1970 by the National Academy of Sciences to secure the services of eminent members of appropriate professions in the examination of policy matters pertaining to the health of the public. The Institute acts under the responsibility given to the National Academy of Sciences by its congressional charter to be an adviser to the federal government and, upon its own initiative, to identify issues of medical care, research, and education. Dr. Kenneth I. Shine is president of the Institute of Medicine.

The **National Research Council** was organized by the National Academy of Sciences in 1916 to associate the broad community of science and technology with the Academy's purposes of furthering knowledge and advising the federal government. Functioning in accordance with general policies determined by the Academy, the Council has become the principal operating agency of both the National Academy of Sciences and the National Academy of Engineering in providing services to the government, the public, and the scientific and engineering communities. The Council is administered jointly by both Academies and the Institute of Medicine. Dr. Bruce M. Alberts and Dr. Wm. A. Wulf are chairman and vice chairman, respectively, of the National Research Council.

COMMITTEE TO REVIEW THE ROLE OF PUBLICLY FUNDED AGRICULTURAL RESEARCH ON THE STRUCTURE OF U.S. AGRICULTURE

ANTHONY S. EARL, *Chair*, Quarles & Brady LLP, Madison, Wisconsin
MICHAEL BOEHLJE, Purdue University, West Lafayette, Indiana
R. DEAN BOYD, Pig Improvement Company, Franklin, Kentucky
FREDERICK H. BUTTEL, University of Wisconsin, Madison
ARNOLD DENTON[*] (retired), Campbell Soup Company, Sacramento, California
ESSEX E. FINNEY[†] (retired), Agricultural Research Service, United States Department of Agriculture, Beltsville, Maryland
CORNELIA BUTLER FLORA, Iowa State University, Ames
PETER J. GOLDMARK, DJR Research, Okanogan, Washington
FREDERICK KIRSCHENMANN, Iowa State University, Ames
DAVID ZILBERMAN, University of California, Berkeley

Staff

CLARA COHEN, *Study Director* (since November 2000)
LEE PAULSON, *Study Director* (from August 1999 to November 2000)
MARY JANE LETAW, *Study Director* (from August 1997 to September 1999)
LUCYNA KURTYKA, *Program Officer* (from August 1999 to July 2000)
ANNE H. KELLY, *Editor*
KAREN L. IMHOF, *Project Assistant*
MICHAEL R. KISIELEWSKI, *Project Assistant*

Consultant
FRED C. WHITE, University of Georgia, Athens

---

[*] Resigned, November 4, 1999
[†] Resigned, December 10, 1999

BOARD ON AGRICULTURE AND NATURAL RESOURCES

HARLEY W. MOON, *Chair*, Iowa State University, Ames
CORNELIA B. FLORA, Iowa State University, Ames
ROBERT B. FRIDLEY, University of California, Davis
BARBARA GLENN, Federation of Animal Science Societies, Bethesda, Maryland
W. R. GOMES, University of California, Oakland
LINDA GOLODNER, National Consumers League, Washington, D.C.
PERRY R. HAGENSTEIN, Institute for Forest Analysis, Planning, and Policy, Wayland, Massachusetts
GEORGE R. HALLBERG, The Cadmus Group, Inc., Boston, Massachusetts
CALESTOUS JUMA, Harvard University, Cambridge, Massachusetts
GILBERT A. LEVEILLE, McNeil Consumer Healthcare, Denville, New Jersey
WHITNEY MACMILLAN, Cargill, Inc., Minneapolis, Minnesota (retired)
TERRY L. MEDLEY, DuPont BioSolutions Enterprise, Wilmington, Delaware
WILLIAM L. OGREN, U.S. Department of Agriculture (retired)
ALICE PELL, Cornell University, Ithaca, New York
NANCY J. RACHMAN, Novigen Sciences, Inc., Washington, D.C.
G. EDWARD SCHUH, University of Minnesota, Minneapolis
BRIAN STASKAWICZ, University of California, Berkeley
JOHN W. SUTTIE, University of Wisconsin, Madison
JAMES TUMLINSON, Agriculture Research Service, U.S. Department of Agriculture, Gainesville, Florida
JAMES J. ZUICHES, Washington State University, Pullman

Staff

WARREN MUIR, *Executive Director*
CHARLOTTE KIRK BAER, *Director*
JULIE ANDREWS, *Senior Project Assistant*

Preface

The food and agricultural economy is highly concentrated today. Economic concentration characterizes food distribution and processing, agricultural inputs, and, increasingly, primary production and commercial farming. Six million farms produced the nation's food during World War II. Today, 90 percent of all farm output comes from fewer than a million farms. This trend is unlikely to be reversed, but it nonetheless troubles U.S. society, which values the concept of the family farm, as farm legislation consistently mentions the family farm as part of its justification and goals.

Vertical integration and contracting increasingly characterize the U.S. food and agricultural system. Vertically integrated farming, processing, and marketing activities often are components of a single corporate entity. Subcontractors might manage a crop or livestock operation while livestock and other assets are owned and much of the decision making is controlled by the farm, which acts as integrator.

In contrast to vertical integration and contracting, but also in response to highly differentiated consumer demands, is the rise in some regions of a segment of farmers engaged in production for niche markets. Niche marketers produce specialty crops or use alternative management practices and typically are independent, small-scale producers. They often market directly to small grocers, specialized outlets, or urban farmers' markets.

The changing food and agricultural system poses major challenges for the public sector's food and agricultural research and educational system. One major challenge is the complexity associated with meeting the research, extension, and education needs of agricultural producers. A concern expressed by Congress and other observers is the putative role of publicly funded agricultural research in developing technologies that have been or will be biased toward changes in farm size and other characteristics of the structure of agriculture.

In response to these concerns, the U.S. Department of Agriculture (USDA) asked the National Research Council (NRC) to review the relationship between publicly funded research and the evolving structure of agriculture.

The NRC convened a study panel of experts chosen for their knowledge of agricultural policy issues, farm and agribusiness management and finance, rural sociology, agricultural economics, and the land grant system. The committee had the following charges:

- Assess the role of public-sector research on changes in farm size and numbers, with particular emphasis on very-large-scale operations.
- Review relevant literature, including pertinent rural development literature, on the role of research and the development of new technology in promoting structural change in farming, evaluating theoretical and empirical evidence.
- Consider whether public-sector research has influenced the size of farm operations and, if so, by what means.
- Provide recommendations for future research and extension policies, giving consideration to improving access to the results of public-sector research that leads to new farm production practices and technology.

As part of its information-gathering activities, the committee held public workshops to elicit the perspective of producers, particularly those who are often described as underserved by the current public research agenda, and other experts on the structure of agriculture. The committee reviewed a wide array of background material, including long-term trends in public and private agricultural research, USDA budgets, literature on economic and sociologic research, literature on adoption and diffusion, and on the relationship among public research, farm size, and structural characteristics. The committee also considered reports on minority and women farmers, the report of the USDA National Commission on Small Farms, and reports by the Congressional Office of Technology Assessment on the structure of agriculture.

This report analyzes the implications of public-sector research, technology adoption, technology transfer, and distribution of public-sector research investments for the structure of agriculture. The report also frames public-sector research and development in the context of other drivers of structural change in agriculture, including market forces, public policy, and the changing role of knowledge and information. The study committee offers recommendations for changes in the public sector's research approach and priority-setting process and for strengthening research programs devoted to analysis of structural change, its causes, and its implications. The study committee hopes that Congress and the Executive Branch will use these recommendations ultimately to benefit a broad diversity of agricultural constituents.

Anthony S. Earl, Chair
Committee to Review the Role of Publicly Funded Agricultural Research on the Structure of U.S. Agriculture

Acknowledgments

Many individuals generously contributed their time, advice, data, and other input during the study process. The committee gives special thanks to those who provided input at its public workshops:

    GLENN ANDERSON, Organic Farms
    ROBERT EVENSON, Yale University
    GAIL FEENSTRA, University of California, Davis
    ROGER GERRITS, (formerly) Agricultural Research
        Service, U.S. Department of Agriculture
    ROBERT GOODMAN, University of Wisconsin, Madison
    CHARLES HASSEBROOK, Center for Rural Affairs
    JAY HARMON, Iowa State University
    DOUGLAS JACKSON-SMITH, Utah State University
    DESMOND JOLLEY, University of California, Davis
    MARLYN JORGENSEN, Jorg-Anna Farms, Iowa
    NOEL KEEN, University of California, Riverside
    EDWARD KNIPLING, Agricultural Research Service, U.S.
        Department of Agriculture
    MARGARET KROME, Michael Fields Agricultural Institute
    MICHELLE MASCARENHAS, Occidental College
    DANIEL MOUNTJOY, Natural Resources Conservation Service,
        U.S. Department of Agriculture

TERRY NIPP, AESOP Enterprises, Ltd.
KEN OLSON, American Farm Bureau Federation
CALVIN QUALSET, University of California, Davis
SARAH J. ROCKEY, Cooperative State Research, Education, and Extension Service, U.S. Department of Agriculture
JAMES VAN DER POL, Minnesota Institute for Sustainable Agriculture
MICHAEL WEHLER, Upland Prairie Farms, Wisconsin
JAMES ZUICHES, Washington State University

The following are also acknowledged for assisting the National Research Council staff during preparation of the report by providing information and statistics: George Cooper, Cooperative State Research, Education, and Extension Service, U.S. Department of Agriculture; Robert Hoppe, Economic Research Service, U.S. Department of Agriculture; Charles Krueger, Pennylvania State University; Richard Pirog, Leopold Center for Sustainable Agriculture; Deborah Sheely, Cooperative State Research, Education, and Extension Service, U.S. Department of Agriculture; Frank Shelton, National Agricultural Statistics Service, U.S. Department of Agriculture; Barbara Tidwell, National Agricultural Statistics Service, U.S. Department of Agriculture; Luther Tweeten, Oklahoma State University; and Dennis Unglesbee, Cooperative State Research, Education, and Extension Service, U.S. Department of Agriculture.

The committee is extremely grateful to the staff members of the National Research Council Board on Agriculture and Natural Resources (BANR) for their efforts throughout the study process and the preparation of this report. The committee also would like to acknowledge Fred C. White for technical assistance, Anne H. (Kate) Kelly for her editorial assistance, and Michael Kisielewski for his efforts in preparing the report. The committee wishes to extend special thanks to Clara Cohen, Study Director since November, 2000, for her dedicated efforts in shepherding the completion of the report.

This report has been reviewed in draft form by individuals chosen for their diverse perspectives and technical expertise, in accordance with procedures approved by the NRC's Report Review Committee. The purpose of this independent review is to provide candid and critical comments that will assist the institution in making its published report as sound as possible and to ensure that the report meets institutional standards for objectivity, evidence, and responsiveness to the study charge. The review comments and draft manuscript remain confidential to protect the integrity of the deliberative process. We wish to thank the following individuals for their review of this report:

MAGGIE ADAMEK, Vision for Change, University of Minnesota
JANET BOKEMEIER, Michigan State University
RUSSELL CROSS, Future Beef

ACKNOWLEDGMENTS

WALLACE HARDIE, Fairmount, North Dakota
JAY HARMON, Iowa State University
CHARLES HASSEBROOK, Center for Rural Affairs
WALLACE HUFFMAN, Iowa State University
D. GALE JOHNSON, University of Chicago
NEAL VAN ALFEN, University of California, Davis

Although the reviewers listed above have provided many constructive comments and suggestions, they were not asked to endorse the conclusions or recommendations nor did they see the final draft of the report before its release. The review of this report was overseen by James Cook, Washington State University, and W. R. Gomes, University of California, Oakland. Appointed by the National Research Council, they were responsible for making certain that an independent examination of this report was carried out in accordance with institutional procedures and that all review comments were carefully considered. Responsibility for the final content of this report rests entirely with the authoring committee and the institution.

Contents

EXECUTIVE SUMMARY 1
    The Study Process, 2
    Project Scope, 3
        Publicly Funded Agricultural Research, 3
        The Structure of Agriculture, 4
    Conclusions and Findings, 5
    Recommendations, 8
        Research Approach, 8
            Broaden Public Goals Beyond Production and Efficiency, 8
            Biophysical and Social Sciences Research, 9
            Public Research, Stakeholder Participation, and
                Accountability, 10
            Assess the Structural Impacts of Publicly Funded Agricultural
                Research, 11
        Extension Policy, 11
            Respond to Broad Variety of Producers, Particularly
                Underserved Populations, 11
            Extension and Engagement, 13
        Future Research, 13
            Monitor and Analyze Structural Change, 13
            Serve Diverse Producers, 14

1 INTRODUCTION	16

The Study Process, 22
Structural Changes in U.S. Agriculture, 24
Publicly Funded Agricultural Research, 26
 Sources of Public Resource Funds, 26
 Institutions Performing Publicly Funded Agricultural Research, 27
Privately Funded Agricultural Research, 28
Report Organization, 29

2 STRUCTURAL IMPACTS OF RESEARCH	30

Research and the Structure of Agriculture, 30
Innovation and the Structure of Agriculture, 32
 Mechanical Innovations, 33
 Chemical Innovations, 34
 Innovations in Biology, 35
 Managerial Innovations, 37
Innovations Applied, 39
 Green Revolution, 39
 Tomato Harvester, 40
 Animal Agriculture, 41
Structural Implications of the Research Priority-Setting Process, 43
 Criteria for Setting Priorities in Agricultural Research, 43
 Stakeholder Participation, 48
 Structural Impact Assessments, 49
Summary, 50

3 STRUCTURAL IMPLICATIONS OF TECHNOLOGY TRANSFER AND ADOPTION	52

Factors that Affect Technology Adoption, 52
 Farm Size, 53
 Regional Differences in Land Quality, 55
 Human Capital, 56
 Producer Age, 56
 Tenure Arrangements, 57
Responding to a Broad Variety of Producers, Including Underserved
 Populations, 57
Technology Transfer, 59
 Market-Oriented Technology Transfer, 60
 Extension: Public-Sector Technology Transfer, 61
Structural Impacts of Extension, 61
 Populations Targeted by Extension, 61

Changing the Focus of Technology Transfer Programs, 63
  Partnerships with the Private Sector, 63
  Partnerships among Public-Sector Institutions, 64
    Changes in Extension Structure, 65
    Changes in Extension Function, 65
    Changes in Extension Process, 67
Summary, 68

4 STRUCTURAL IMPACTS OF PUBLIC INVESTMENT IN AGRICULTURAL RESEARCH 69

Public-Sector Responses to Structural Issues, 69
Agricultural Research Investments, 72
  Current Research Information System, 72
  Public Research Spending, 1986 and 1997, 73
  Public Research Spending, 1999, 78
Public Research and Environmentally Sustainable Alternative Agriculture, 79
Structural Implications of Research Funding Mechanisms, 82
  Fund for Rural America, 82
  Initiative for Future Agriculture and Food Systems, 83
  National Research Initiative Competitive Grants Program, 83
Summary, 85

5 DRIVERS OF STRUCTURAL CHANGE, CHANGES IN KNOWLEDGE AND INFORMATION, IMPLICATIONS FOR POLICY 86

Drivers of Structural Change, 86
  Relative Price of Labor and Capital, 87
  Knowledge and Information: A Changing Role, 89
  Government Policy and Structure, 90
Changes in Farming, 92
  Global Competition, 92
  Industrialized Agriculture, 92
  Differentiated Products, 93
  Precision (Information-Intensive) Production, 93
  Ecologic Agriculture, 94
  Food Supply Chains, 94
  Increasing Risk, 95
  Increasing Diversity, 96
Information, Innovation, and the Structure of Agriculture, 97
  Structure and Coordination, 97
  Intellectual Property Rights and Distributional Consequences, 99
  Global R&D and Information, 101

Access to Technology and Disenfranchisement, 101
Research Opportunities, 103
Summary, 103

REFERENCES   105

APPENDIXES
A Committee to Review the Role of Publicly Funded Agricultural
   Research on the Structure of U.S. Agriculture, Public Workshop,
   November 19, 1999   119

B Committee to Review the Role of Publicly Funded Agricultural
   Research on the Structure of U.S. Agriculture, Public Workshop,
   January 18, 2000   122

C U.S. Public (USDA and State Agricultural Experiment Stations)
   and Private Agricultural Research Funds by Performing Organization,
   1888–1990 (Millions of Dollars)   125

D Economic Research Service Farm Typology   129

ABOUT THE AUTHORS   131

BOARD ON AGRICULTURE AND NATURAL RESOURCES
   PUBLICATIONS   135

Tables, Figures, and Boxes

Tables

3-1 Contact with Extension by Herd Size, 62
4-1 Historic Allocation of Public Research Funds by Commodity, 74
4-2 Historic Allocation of Public Research Funds by Goal, 75
4-3 Allocation of Public Agricultural Research Funds, 1999, 77
4-4 Selected Alternative Agricultural Technologies: Current USDA-Funded Projects and Total Patents Granted, 1975–1998, by Type of Organization, 80

Figures

1-1 Number of Farms and Acres Per Farm, 1850–1997, 17
1-2 Smallest Percentage of U.S. Farms Accounting for Half of the Nation's Agricultural Sales, Selected Years from 1900–1997, 18
1-3 Share of Hog Production by Type of Vertical Coordination, 1970–1999, 20
3-1 Technology Adoption by Dairy Herd Size, 1999, 54

Boxes

1-1 Public Sources of Funding for Agricultural Research, 26

3-1 Responding to Asian Growers in California, 59
3-2 State Extension Partners are Linking to Other Federal Agencies on a Broad Array of Problems, 66
3-3 Stakeholder Participation and SARE, 67
3-4 Fax-Based, Satellite Information Request System: Reaching Small and Part-Time Farmers, 68
4-1 Public-Sector Responses to Structural Issues, 70
4-2 Research Funding and Structural Change, 83

Executive Summary

The U.S. food and agricultural sector is undergoing rapid change in production, distribution, and consumption of food and fiber, and in technology. There have been dramatic increases in production and marketing coordination, market contracting, concentration of agricultural output by fewer and fewer operations, and consolidation of agricultural operations. These increases are manifested in significant long- and short-term changes in farm size, number, distribution, and location. Production that once relied on small, independent, family-based farms increasingly occurs in large, consolidated, global operations. Small- and mid-sized operators often struggle to remain competitive and to adopt recent developments in technology and information.

The changes occurring in the modern food and agricultural system pose major challenges for public-sector agricultural research and education. One challenge is the complexity of serving and meeting the needs of agricultural producers—both the large commercial agricultural production sector and the multitude of smaller producers, including low-income and limited-resource producers, and producers of niche commodities. There is concern that publicly funded agricultural research has influenced the development of technologies that have been or will be biased toward changes in farm size and industrialization of the farm sector. There is debate about whether publicly funded agricultural research is equally accessible to all users and whether it is targeted to the full range of user and citizens' groups.

This report analyzes publicly funded agricultural research and the structure of agriculture, and it offers recommendations for research and extension policies. It evaluates the applicability of publicly funded agricultural research across the agricultural distribution system: from small, poorly capitalized farms to large, well-capitalized industrial organizations. Although the committee acknowledges that the public sector has been encouraged, and in some cases mandated, to serve constituents, as illustrated by the increasing public policy support for small farmers and other underserved groups in the last four farm bills, the focus of this report is on analysis without judgment about the social desirability of particular distributions.

THE STUDY PROCESS

The U.S. Department of Agriculture (USDA) requested that the Board on Agriculture and Natural Resources of the National Research Council (NRC) convene a panel of experts to examine whether publicly funded agricultural research has influenced the structure of U.S. agriculture and, if so, how. The Committee to Review the Role of Publicly Funded Agricultural Research on the Structure of U.S. Agriculture was asked to assess the role of public-sector agricultural research on changes in the size and numbers of farms, with particular emphasis on the evolution of very-large-scale operations. The committee's charge was as follows:

- Review relevant literature, including pertinent rural development literature, on the role of research and the development of new technology in promoting structural change in farming, evaluating theoretical and empirical evidence.
- Consider whether public-sector research has influenced the size of farm operations and, if so, by what means.
- Provide recommendations for future research and extension policies, giving consideration to improving access to the results of public-sector research that leads to new farm production practices and technology.

The committee analyzed publicly funded agricultural research documented in the Current Research Information Systems (CRIS) database, which is the USDA's documentation and reporting system for research projects in agriculture, food and nutrition, and forestry. It also considered information drawn from case studies and from the scientific literature. The committee gathered input and information from stakeholders during two public workshops held in conjunction with this study (Appendixes A and B).

PROJECT SCOPE

Publicly Funded Agricultural Research

Publicly funded agricultural research comprises a complex variety of programs, users, and funding sources. The committee considered *publicly funded agricultural research* to be any agricultural research performed with financial and material support from the public sector. Sources of public-sector support include international organizations and federal, state, and local governments. The proportion of public funds in any research activity varies by institution and project. Publicly funded agricultural research is performed in public- and private-sector institutions.

The committee elected not to survey and analyze comprehensively all sources of publicly funded agricultural research, given the challenges in defining and disaggregating investments in agricultural research over time across different agencies and in determining their relationship to structural variables. The committee instead chose to limit the scope of its analysis to a subset of publicly funded agricultural research that could be used as a proxy for the wider scope of research described above. The committee chose to emphasize the principal components of USDA-supported agricultural research and extension, including extramural scientific research support to state-level partners and other research programs administered by the Cooperative State Research, Education, and Extension Service (CSREES); intramural biophysical science research conducted by the Agricultural Research Service (ARS); intramural social science research conducted by the Economic Research Service (ERS); the collection and analysis of agricultural data by the National Agricultural Statistics Service (NASS); and research conducted by the USDA Forest Service. The committee also considered state- and federally supported institutions, including land grant institutions, colleges of agriculture, agricultural experiment stations, the Cooperative Extension service, schools of forestry, historically African American land grant institutions, colleges of veterinary medicine, colleges of human sciences, Native American land grant institutions, and Hispanic-serving institutions.

The committee relied heavily on data reported to the CRIS database by USDA intramural research agencies, state agricultural experiment stations, 1890 and 1862 land grant universities, state schools of forestry, schools of veterinary medicine, and USDA grant recipients. Although the committee acknowledges limitations of the CRIS in comprehensively reporting research conducted by other non-USDA agencies, CRIS is the most reliable and consistent database available.

The committee acknowledges that private sources of funding for agricultural research have grown significantly relative to public sources

(Huffman and Evenson, 1993; See table in Appendix C). The committee recognizes that mutual influence exists between private and public research, both through the input of public research results into private-sector research and through the influence of the private sector on the public-sector research agenda through funding provision. Public-private partnerships are increasingly used as a mechanism for technology transfer (discussed in greater detail in Chapter 3). The committee acknowledges that linkages between the public and the private sectors are important and are likely to have significant implications for the structure of agriculture. However, an analysis of the relationship between privately funded agricultural research and the structure of agriculture is beyond the scope of this report. It is an important issue that should be the focal point of further analysis.

The Structure of Agriculture

The "structure of agriculture" is a broad phenomenon involving both the characteristics of farms (the system of agricultural production) and the relationships of farms to other sectors and institutions. "Structural change" refers to the change in those characteristics over time. While there is no universally accepted definition, there is considerable agreement that farm structure involves matters such as:

- The size (measured in acreage or gross farm sales) and size distribution of agricultural operations, including the concentration of agricultural production—the increasing share of agricultural output by fewer firms.
- The number of agricultural operations.
- The spatial character of production systems.
- The technology and production characteristics of agricultural operations, including the level of specialization and diversification.
- Resource ownership arrangements, including tenancy and leasing.
- The relationship among ownership, management, and labor.
- Dependence on primary resources, or the relative degree to which an operation depends on capital, labor, or knowledge.
- Inter- and intrasectoral linkages, including contractual relationships for marketing and production.
- The extent and pattern of vertical and horizontal integration.
- Production, marketing, and financial management strategies.
- Business organization (including sole proprietorships, partnerships, corporations and cooperatives) and arrangements (including joint ventures, leasing, independent production, and contracting).

- Characterization of the workforce, including age, gender, ethnicity, education, experience, skill level, and part-time versus full-time status of the operator.

In recent years, the dimensions of agricultural structure of greatest interest in public policy discussions have been farm numbers, the size distribution of farms (and concentration of farm production), and relationships of vertical integration among input providers, farm producers, and agricultural processors. This report will emphasize these structural characteristics but will also consider other dimensions of structure such as the spatial distribution of production, spatial specialization of production, and part-time farming. There is very little comprehensive, quantitative research on the relationship of research to many structural variables.

CONCLUSIONS AND FINDINGS

A focused analysis of the first two portions of the committee's charge leads to the following conclusions:

Conclusion 1: Public-sector agricultural research is an important, but not an exclusive, factor in structural change. The commodity and production orientations of public-sector agricultural research have contributed to concentration in the industry.

Innovation leads to change, and, almost invariably, that change has a structural dimension. Agricultural research, including publicly funded efforts, will influence structural change. Very little empirical evidence exists on the effects of publicly funded agricultural research on structural variables. What little exists demonstrates that the amount of and rate of change in publicly financed agricultural research and development and education are correlated with increases in average farm size, with the number of very large farms (1,000+ acres), with large farms as a percentage of all farms, with livestock specialization, and with farmers' off-farm work participation (Busch et al., 1984; Huffman and Evenson, 2001). A detailed analysis of individual research areas indicates that, although some research areas, such as those involving mechanical innovation, are more likely to encourage concentration, many others (biologic, chemical, and managerial innovations and innovations that address environmental issues) can have mixed effects on structure. The overall relationship between publicly funded agricultural research and structural change is not clearly defined.

Significant distributional and structural changes are associated with factors other than publicly funded agricultural research, including market forces, public policy, and the changing role of knowledge and information. Market forces, including changes in the availability of financial capital, changes in international capital flows, and the increasing price of labor relative to capital tend to encourage expansion of larger, capital-intensive firms (Cochrane, 1979; Hayami and Ruttan, 1985; Kislev and Peterson, 1981; 1982; 1996). Factors that lower the profit margin per unit of product also tend to increase the size of production units.

Public policies other than for research and development also can influence structure. Commodity payment policy, crop insurance policy, conservation policies, farm loan policies, federal income and estate tax law, and labor and environmental regulatory policies can have significant structural implications (Carman, 1997; Durst and Monke, 2001; GAO, 2001; Goetz and Debrtin, in press; Lichtenberg et al., 1988; Sisson, 1982; USDA, 1998a; Williams-Derry and Cook, 2000; Zilberman, et al., 1991).

Knowledge and information are also becoming increasingly important drivers of control and structural change in the agricultural industry, and access to information and intellectual property rights are becoming sources of conflict and controversy as the value of information increases and as that value can be captured by the private sector.

The framing of agricultural research within the context of these other drivers demonstrates that distributional effects result from many factors (Chapter 5). Structural change should not surprise us—it is a consequence of market forces, public policy, and other factors at work in a commodity industry.

Conclusion 2: Public-sector research and technology transfer are not always scale neutral; thus, different groups adopt research results disproportionately.

The ability of agricultural operations to make use of research is unequal because of differences in farm size; regional land quality; and the age, wealth, education, access to credit, and human capital resources of the operators. Heterogeneity among producers influences what technology is adopted, to what extent, and when. Larger operators who have access to sufficient human resources and financial capital are more likely than smaller operators to use the products of research, including scale-neutral technologies. Smaller operators may take more time to adopt technologies (Just and Zilberman, 1988). Thus, by developing new technologies and introducing change throughout the agricultural system, public research can favor dynamic, usually larger farmers. Given the perceived-risk, fixed-cost, and credit constraints of adopting new technology, small- and mid-sized operators often find it difficult to compete with larger businesses.

Conclusion 3: Technology transfer is a factor in structural change.

Cooperative Extension, the technology transfer arm of the public sector, is an important link between research and the structure of agriculture. A major function of the extension service has been to communicate research results to farmers and other citizens through adult education. Extension can affect agricultural structure through *what* is communicated, to *whom,* and *how* it is communicated. Evidence suggests that extension works disproportionately with some groups: by race and gender of the operator, by the size of the farm, and by the type of agriculture (Flora et al., 1993; Ostrom et al., 2000).

Conclusion 4: Publicly funded research is important to the public good. Public-sector agricultural research institutions are beginning to shift their focus to a broader research agenda that supports the production of public goods.

Substantial research funding has been reallocated to goals other than productivity that contribute to the production of public goods—goods from which revenue cannot be captured. Environmental issues, sustainable production systems, resource conservation, and rural development have gained importance. The committee found that a modest but growing share of resources is now allocated to research on techniques and technologies that are of interest to small-scale farmers, organic farmers, and others outside the commercial mainstream. This broader research portfolio is likely to benefit constituents in a variety of circumstances.

Conclusion 5: Public-sector agricultural research is an important element of an integrated policy for addressing distributional inequities. Although distributional issues increasingly are becoming a focal point of publicly funded agricultural research, it is unlikely that changes in public-sector research policy would completely offset or neutralize distributional inequities, given that other forces also encourage structural change.

Public research and development are critical to promoting innovation in and the maintenance of a vibrant agricultural industry. Public support can ensure the development of new research paradigms, technology, and structures; encourage broad access to research results; and secure stakeholder participation in the research effort. The public sector is critical because it can acknowledge distributional issues and target underserved populations that often are overlooked by private-sector research and development.

Public agricultural research is changing to serve and engage input from a variety of stakeholders at the same time that it continues to serve its traditional clientele. Structural and distributional issues have been key areas of legislation, such as the Federal Agriculture Improvement and Reform (FAIR) Act of 1996 (U.S. Congress, 1996), and the Agricultural Research, Extension, and Education Reform Act (AREERA) of 1998 (U.S. Congress, 1998). Public funding for structural and distributional research, including research that monitors structural change as well as that designed to target the needs of specific constituencies, has begun to increase in intramural and extramural public agricultural research programs.

Numerous factors other than public-sector research and development and extension, including market forces and government policies, also promote structural and distributional change. Thus, modification of public research and development policy alone would not be sufficient to offset changes.

RECOMMENDATIONS

A public-policy approach is provided that responds to the committee's third charge, to develop recommendations for future research and extension policies, giving consideration to improving access to results of public-sector research that leads to new farm production practices and technology.

Recommendations are provided in three categories. The first relates to research approaches and broad guidelines for research decision making and priority setting. A second group relates to extension policy. The third category provides a public-policy response for monitoring and analyzing structural change and its causes. Specific opportunities are identified for research in the area of structural change in Chapter 5.

There is a legitimate concern that adoption of the committee's recommendations might result in reduced economic surplus in the aggregate. However, the committee submits that how losses or gains are distributed is also an important question. This issue of distribution among participants in the agricultural industry is a critical and largely missing component of the public-policy debate concerning R&D and innovation investments.

Research Approach

Broaden Public Research Goals Beyond Production and Efficiency

Publicly funded research and development in agriculture historically have emphasized production of commodity products and, over time, production of these commodities for a global market. The committee noted, however, that the public agricultural research system has an obligation to attain multiple objectives, including, but not limited to, the traditional mission of productivity improvement in commodity products sectors, the provision of public goods, and service to a diverse group of constituents who are entitled to access to the fruits of the public research system. Although the publicly funded research agenda has increasingly emphasized areas in addition to commodity production, a broader package of research and development activities might more effectively serve a diverse range of stakeholders who are unlikely to be competitive in commodity markets and who are not served by the private sector. This package could include a greater focus on improving farm income through production of higher value products and on improving farm management to reduce capital expenditures. The committee recognizes that there are limits to the degree to which developing technology for "niches" is sustainable, since increased research and development on a niche product will increase the size of the market, invite entry by other producers, and thereby turn the niche product into a commodity product. The quest for higher value niche production technology and products is thus a perpetual one. Nevertheless, a more broadly defined publicly funded research agenda could serve an increasingly diverse industry that includes small-scale producers, producers using ecologically based agricultural practices, and others outside the commercial mainstream. The committee believes research should be conducted to serve those constituencies.

Recommendation 1
The goals of public-sector research should continue to be broadened beyond productivity and efficiency. Federal and state research should improve technology and information systems that benefit farmers in diverse production systems and circumstances, including part-time farmers, small-scale farmers, organic farmers, and value-added producers. However, limiting public-sector research to scale-neutral technologies is not sufficient to meet the needs of a diverse producer constituency. The public sector increasingly should assess the opportunities for R&D and technology transfer for those who are not served by the private sector.

Biophysical and Social Sciences Research

Developing an interdisciplinary research approach that integrates social science and humanities perspectives is critical to setting priorities and to understanding the relationships that influence structural change. Examples from international and domestic contexts illustrate how integration of social science approaches has broadened the research agenda to serve constituents, particularly smaller growers, more effectively (Feldstein and Poats, 1989; Mountjoy, 2001; Rhoades and Booth, 1982). The committee envisions an important role for social science research on agribusiness and entrepreneurial enterprises other than farm management; the costs, benefits, and consequences of technology, including social, human, and community factors; and rural development, including lifestyles and opportunities for individuals and communities. Social science research on farm structure and production systems also can contribute to establishing a needs-assessment baseline in research decision making.

Recommendation 2
The public sector should use an interdisciplinary approach integrating biophysical science, social science, and humanities perspectives to determine structural consequences of research and to assess the research needs of a diverse clientele. The public sector, particularly ARS, should strengthen social science expertise in the areas of setting research priorities and assessing the distributional implications of research and new technology.

Public Research, Stakeholder Participation, and Accountability

The committee suggests that publicly funded agricultural research should be more accountable to the public, and it endorses public participation as a vital step in ensuring that diverse stakeholder needs are met through public-sector research. Participatory methods have been used successfully in the Consultative Group on International Agricultural Research (CGIAR) and in other international contexts to determine the agenda for plant breeding, crop, and natural resource management research that benefits small-scale agricultural operations. The committee recognizes the public sector's efforts to increase stakeholder participation in decision making. The committee submits, however, that to the degree that public involvement and stakeholder participation in setting research priorities focus on existing commodity groups, this input will not lead to results different from the traditional emphasis of the public research system on developing new technology for commodity production. Public involvement of a broader representation of stakeholders should be promoted, and the process for involving the public analyzed to improve effectiveness and transparency.

Recommendation 3
To improve accountability to constituents, the public sector, at both the federal and state levels, should continue to incorporate the knowledge and needs of stakeholders through genuine public participation in setting priorities for research and in implementing research projects; encourage broad-based participation on research and extension advisory boards to assess the relevance and importance of proposed research and extension programs and to ensure that priority setting is responsive to a variety of needs, particularly those that cannot be met by the private sector; conduct critical analysis and assessment of the methods used for engaging, interpreting, and incorporating

stakeholder input into decision making; and take action to make the participation process more understandable and transparent to the public.

Assess the Structural Impacts of Publicly Funded Agricultural Research

Data and research on the relationship between public research and structural change are limited. *Ex ante* impact assessment research on prospective technologic thrusts and *ex post* research on recently commercialized technologies are most urgent when these technologies are likely to have major impacts on the structure of agriculture, the environment, food safety, or the relations between agriculture and consumers.

Recommendation 4
Public-sector research institutions, at both the federal and state level, should develop expertise and research programs devoted to analyzing the distributional implications and impacts of agricultural R&D for various groups of producers, using both ex ante and ex post research designs. The study committee endorses the public sector's earlier efforts in this regard and encourages continued development of this research base.

Extension Policy

The Smith-Lever Act of 1914 established a role for extension personnel to disseminate useful and practical information to farmers and farm families. Over time, that role has changed, as much of the technology information once provided by extension is now provided by the private sector, particularly suppliers of agricultural inputs such as seeds and fertilizer. In addition, the public now needs information on a greater diversity of issues, such as environmental and public health issues.

Respond to Broad Variety of Producers, Particularly Underserved Populations

The ability of farmers to take advantage of public-sector research results depends on many factors, including farm size; land quality; access to markets, labor, capital, and land; and the race, ethnicity, age, gender, and education of the operators. The public sector should enhance outreach to meet the needs of this heterogeneous group, particularly of those who have not been well served by the current research agenda. The committee acknowledges that some publicly funded research and outreach programs and projects do target specific underserved populations. Examples include the USDA Small Farms Program, the Hispanic-Serving Institutions Educational Grants Program, the Tribal Colleges Research Program, and the 1890 Institutional Capacity Building Program. The committee encourages the public sector to continue to expand programs that fund research on the topics, processes, and audiences represented by these minority-serving institutions. However, the committee also submits that the public-sector response to these populations has been less than proactive and initiated only in response to considerable public pressure and such litigation as the 1997 class action lawsuit filed against USDA by African American farmers (Pigford v. Glickman, 1997).

Many factors that characterize underserved groups, including size of farm, race, or ethnicity, are intangible policy parameters for research decision making. For example, there is such a diversity of small and medium-sized farms that it is difficult to generalize what they share in terms of research needs. It is much easier for a public research system to respond to needs of underserved populations if it can target concrete production systems that have promise and can be funded readily. If coupled with rigorous needs assessment to identify the production systems used by underserved populations, targeting nonmainstream or niche types of production systems can sometimes be a good proxy for reaching underserved populations.

Public-sector outreach activities, including extension, should serve a variety of producers—including limited-resource producers, organic producers, direct-marketing producers, transition farmers, full- and part-time farmers, and cooperatives—with continued special efforts to reach underserved or minority communities.

Recommendation 5
Public-sector outreach, including extension, should take a proactive role in assessing the research and development and technology transfer needs of a variety of producers, including underserved and minority groups; designing appropriate strategies, such as applied on-farm research, for serving those constituencies; and providing production assistance and other appropriate services, such as market development education for differentiated product markets, entrepreneurship education, financial strategies, value-added processing, and identification of opportunities for those working part time in agriculture.

Recommendation 6
The public sector, at both the federal and the state level, should expand its programming focus with minority-serving institutions, which have unique access to underserved groups.

More effective communication with these groups would help research institutions move toward conducting research and extension that are relevant to their circumstances.

Extension and Engagement

The committee endorses new models of engagement that may help extension more effectively serve an increasingly diverse clientele. For example, Cooperative Extension is increasingly forming novel partnerships with the private sector, particularly with regard to the production of public goods. Extension also is developing public-sector partnerships, both within universities and among other federal institutions, to access the expertise needed to respond to an array of problems that go beyond agricultural production or farm programs. Extension is increasingly engaging farmers and others in the research process and improving accessibility to information for many constituencies.

A more broadly defined extension service may ensure greater use of research results and technology by more diverse clientele.

Recommendation 7
Extension should continue to reach out to other programs within universities, to draw wider networks of human resources, and to work with broader arrays of partners in the federal, private, nonprofit, and client sectors. CSREES should continue to facilitate more interdisciplinary and interagency activities involving its state extension partners. CSREES should evaluate the potential and effectiveness of these extension approaches to serve diverse constituents.

Future Research

The study committee envisions research in three major areas: to monitor and analyze structural variables; to serve the needs of diverse constituencies; and to further explain how other factors—including market forces, government policy, and knowledge and information—drive structural change.

Monitor and Analyze Structural Change

The committee acknowledges recent public-sector efforts to monitor and analyze structural variables. ERS, for example, has developed a significant body of research on structural trends, including a new farm classification system to divide U.S. farms into mutually exclusive and more homogeneous groups for more refined analysis (Hoppe et al., 2000). This and other research results are included in a Farm Structure Briefing Room on the ERS website (http://www.ers.usda.gov/briefing/FarmStructure). It is important to note, however, that funding for such programs is still minimal relative to funding allocated to production agriculture. The public sector should continue and expand its efforts to track the changing structure of agriculture.

The other drivers and determinants of structural change are numerous: pressure from consumers and end-use markets, changing demographics and work habits of U.S. families, changing attitudes about food safety and quality, increasing competition from global market participants, economies of size and scope in production and distribution, risk mitigation and management strategies of buyers and suppliers, strategic positioning and market power or control strategies of individual businesses, and private-sector research and development and technology transfer policies. The committee recognizes the public sector's efforts thus far to investigate these driving forces and the effects of alternative policy instruments on structural change. The committee encourages an even greater public-sector commitment to studying the structural effects of these drivers. Three of them—market forces; public policies; and knowledge and information, including the increased privatization and globalization of information and research and development markets—have particularly significant structural effects, and all of them have critical implications for the design of research and development policy.

Recommendation 8
The public sector should continue to acknowledge the importance of structural change in agriculture. ERS and NASS should continue to monitor and analyze structural change and its causes.

Serve Diverse Producers

Public research targeting the needs of diverse constituencies has begun to increase, and the committee acknowledges these efforts. The 2001 Request for Proposals (RFP) for the USDA Fund for Rural America encourages research projects that "help increase farm profitability among small and minority farmers" (Federal Register, 2001a). The 2001 RFP for the USDA Initiative for Future Agricultural and Food Systems also highlights distributional concerns related to the viability and competitiveness of small and medium-sized farms as one of six priority programs (Federal Register, 2001b). ERS has conducted research relevant to niche farmers, including a comprehensive analysis of organic farming and marketing (Fernandez-Cornejo et al., 1998) and an assessment of certified organic acreage by state and by commodity (USDA, 2001c). ARS dedicated the October 1999 issue of *Agricultural Research* Magazine to research projects relevant to small farmers and ranchers (USDA, 1999b). Finally, in 1999, USDA awarded $9.6 million in grants for research, training, and education to implement HACCP (Hazard Analysis and Critical Control Point) and other food safety advancements, of which $1.35 million was targeted specifically to assist small meat-processing plants and small farmers (USDA, 1999d). Nevertheless, the committee argues that funding for such programs is still extremely low relative to that devoted to production agriculture research.

Recommendation 9
The public sector should continue to experiment with research approaches—including multifunctional partnerships that link research and extension, partnerships that link the public sector with the private and nonprofit sectors, multi-state cooperation, and multidisciplinary collaboration—as instruments for serving small farms, minority farmers, and other underserved producers. The public sector should evaluate the potential and effectiveness of these research approaches to serve these constituents.

1

Introduction

The U.S. food and agricultural sector is undergoing rapid change in the production, distribution, and consumption of food and fiber, and in technology. There have been dramatic increases in production and marketing coordination, market contracting, concentration, and consolidation that have been manifested in significant long- and short-term changes in farm size, number, distribution, and location.

Agriculture is becoming increasingly consolidated, as reflected in farm number and size (both in acreage and annual sales). Between 1959 and 1992, the number of farms declined by almost half, but average acreage increased 60 percent and average nominal sales grew tenfold (Sommer et al., 1998). Figure 1-1 shows that between 1935 and 1997, the number of farms has declined from 6.8 million to 1.9 million, with most of the change occurring prior to 1974 and leveling off after 1974 (U.S. Bureau of the Census, 1900–1992; USDA, 1999c). Average farm size also has increased, from 155 acres in 1935 to 487 acres in 1997 (U.S. Bureau of the Census, 1900–1992; USDA, 1999c). Similarly, the increase in average farm size leveled off after 1974.

INTRODUCTION

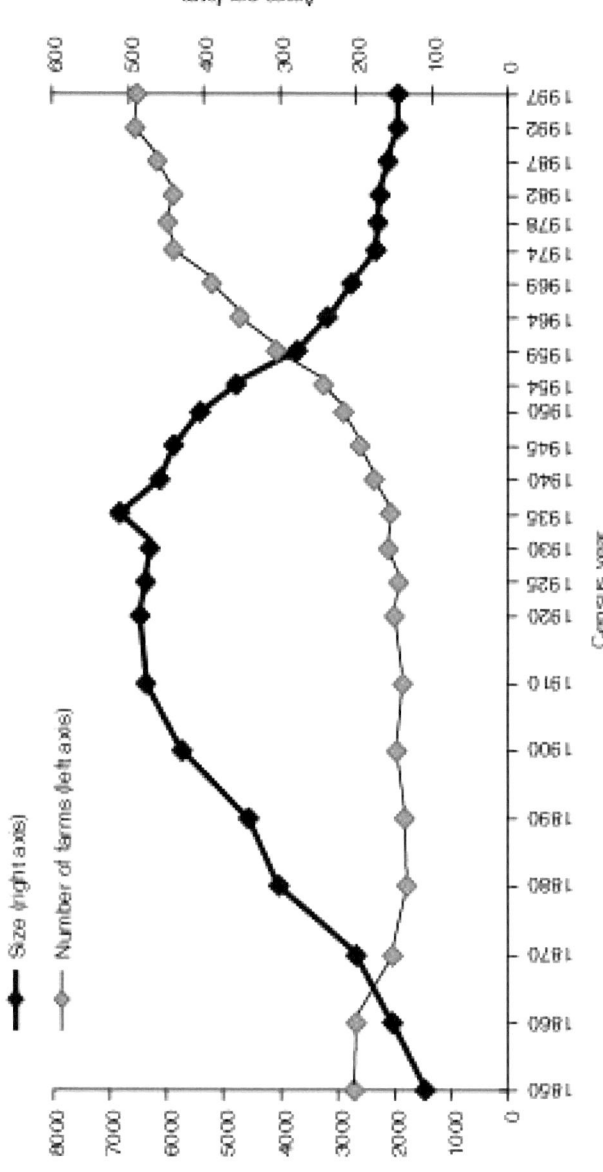

FIGURE 1-1 Number of farms and acres per farm, 1850–1997.

SOURCE: Structural and Financial Characteristics of U.S. Farms: 2001 Family Farm Report, 2001. R. A. Hoppe, ed., Economic Research Service, U.S. Department of Agriculture; USDA, 2001f, based on U.S. Bureau of the Census, 1900–1992 Censuses of Agriculture; U.S. Department of Agriculture, National Agricultural Statistics Service, 1997 Census of Agriculture.

Concentration, as reflected in the increasing share of agricultural output by fewer and fewer farms, also has increased over the past century. Figure 1-2 shows that the smallest percentage of U.S. farms accounting for half of the nation's agricultural sales declined from 17 percent in 1900 to two percent in 1997 (U.S. Bureau of the Census, 1900–1992; USDA 1999c). Particularly dramatic concentration has occurred in the livestock industry. In 1995, four companies controlled more than 80 percent of U.S. beef cattle slaughter (USDA, 1997c), and it is predicted that 40 or fewer agricultural supply and distribution chains will soon dominate the swine industry (Drabenstott, 1998). Although the increased concentration of production in farming has been dramatic, farming is not as concentrated as other industries. Concentration is most dramatic for the industries with which farmers do business (MacDonald et al., 1999; Stanton, 1993).

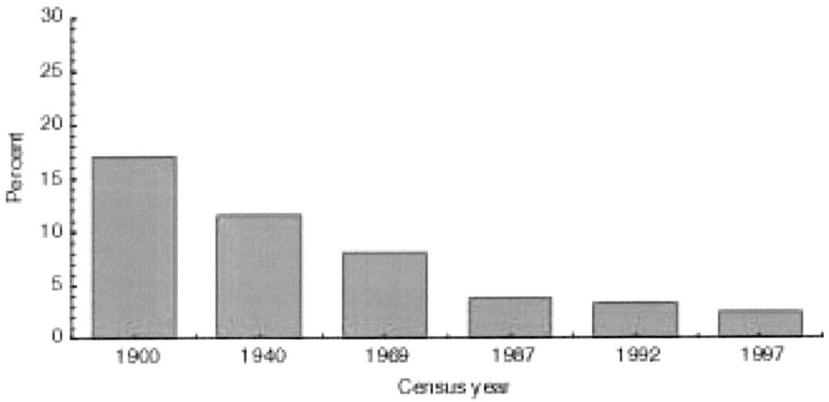

FIGURE 1-2 Smallest percentage of U.S. farms accounting for half of the Nation's agricultural sales, selected years from 1900–1997.

NOTE: The share of sales in 1909, 1940, and 1969 was calculated by summing share of sales by sales class from Census data, and totaled slightly over 50 percent. The share of sales in 1987, 1992, and 1997 was calculated by the Census Bureau using farm-level data and therefore totaled exactly 50 percent.

SOURCE: U.S. Department of Agriculture, Economic Research Service, based on U.S. Bureau of the Census, 1900–1992 Censuses of Agriculture; U.S. Department of Agriculture, National Agricultural Statistics Service, 1997 Census of Agriculture.

Agricultural production is highly concentrated among large family farms (annual sales between $250,000 and $499,999), very large family farms ($500,000 or more), and nonfamily farms (farms organized as nonfamily corporations or cooperatives and farms with hired managers). In 1999, these groups constituted only 8 percent of the total number of farms in the United States, but they accounted for 68 percent of the value of production. In contrast, farms with sales of less than $250000 accounted for 92 percent of all farms but only 32 percent of total agricultural production (USDA, 1999a). Small farms with gross sales less than $250,000 collectively held 72 percent of farm assets, including 74 percent of the land (measured in acres) owned by farms and 87 percent of the land in the Conservation Reserve Program or Wetlands Reserve Programs (USDA, 1999a).

There are also marked distributional differences in farm ownership. In 1999, most agricultural operations were fully owned; part owners accounted for more than half the value of production. Farms with annual sales of less than $10,000 were most likely to be sole proprietorships (USDA, 1999a).

Contracting and vertical integration are increasing in U.S. agriculture. Contracting—agreements between producers and companies or other farmers that specify conditions of production or marketing—has increased from 1960 to the present (Perry and Banker, 2000). Although only 10 percent of U.S. farms reported marketing and production contracts, the combined contract and non-contract production on those farms accounted for about 52 percent of the total value of farm production in 1999 (USDA, 1999a). Vertical integration—coordination of stages in the agricultural product chain under common ownership—increasingly occurs through market or production chains. In vertically integrated operations, the same firm typically owns several farm-related businesses, such as hatcheries, feed mills, processing plants, and packing facilities. The integrator may also own farms, but more typically contracts with farms to produce commodities. Figure 1-3 shows rapid increases in contracting and vertical integration in the hog industry.

One of the most important, but less well recognized, dimensions of structural change in American agriculture has been the regional restructuring of production and of related processing and manufacturing activities. The most dramatic regional shifts in production during the post-World War II period were in the major livestock sectors. In the 1950s through the 1970s fed-cattle production shifted dramatically from the Eastern Corn Belt to the Western Corn Belt and the Southern Great Plains as the expansion of groundwater irrigation enabled rapid growth of feed-grain production in these areas. Broiler production

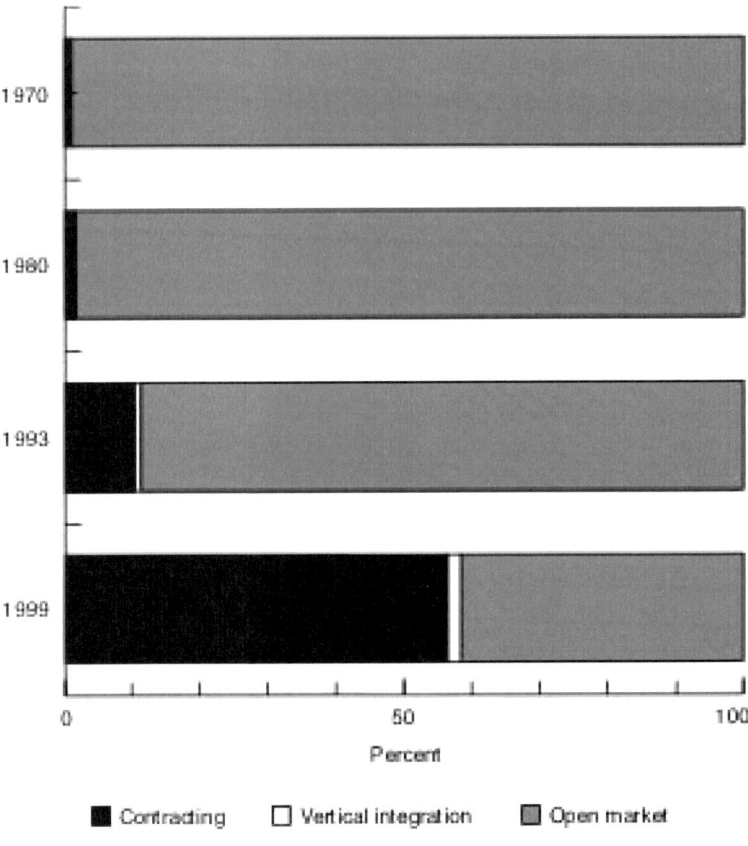

FIGURE 1-3 Share of hog production by type of vertical coordination, 1970–1999.

SOURCE: Compiled by U.S. Department of Agriculture Economic Research Service from various studies. See S. W. Martinez, 1999. Vertical Coordination in the Pork and Broiler Industries. AER-777, U.S. Department of Agriculture, Economic Research Service.

also exhibited rapid shifts and became increasingly concentrated in particular regions (e.g., the Delmarva peninsula, the Mississippi Delta).

More recently, there have been major shifts in the location of dairy and pork production. Manchester and Blayney (1997), for example, have documented rapid shifts in the location of dairy production from 1975 to 1995. The share of national dairy production in the two traditional dairy regions (the Northeast and Lake States), for example, decreased from 48.4 percent in 1975 to 42.8 percent in 1995, while the Mountain and Pacific states witnessed an increase in their share of national dairy production from 16.7 to 29.9 percent over this period of time. Over the past twenty years there was simultaneous growth of pork production in the Southeast, particularly North Carolina, and decreased pork production in the Eastern Corn Belt (McBride, 1997; MacDonald et al., 1999).

Changes in the location and size of meatpacking plants are also associated with regional reorganization of production. MacDonald et al. (1999) have noted that in the case of fed-cattle production, packing plants have tended to relocate (mainly to Nebraska, Kansas, eastern Colorado, and the Texas Panhandle) in response to changes in the location of production. By contrast, in the case of hog production, the evidence suggests that the regional relocation of production has been substantially driven by processors as they have moved to the Southeast in search of cheaper labor and less stringent environmental regulations (Khu and Durrenberger, 1998).

Operator characteristics also have changed. Since the 1930s, farmers have combined off-farm work with farming, but since the 1970s, more than one third have held down full-time jobs off the farm (U.S. Bureau of the Census, 1930–1992; USDA, 1999c). About 90 percent of farm household income came from off-farm sources in 1999, and net earnings from farming activities averaged only $6,400 per farm household (USDA, 1999a). Sources and levels of income vary with farm size. In 1999, average farm earnings per household increased with sales class. On average, the households of small farms depended heavily on off-farm income, while the households of larger farms depended mostly on farm income (USDA, 1999a).

Structural changes in agriculture also are reflected in the changing makeup of what constitutes an underserved farmer group. For example, the number of U.S. farms operated by nonwhite minorities declined from 15 percent of all farms in 1920 to 2.5 percent of all farms in 1997. In 1920, 14 percent of U.S. farms were farms operated by African Americans, but by 1997 only 1 percent was operated by African Americans, and the numbers continue to decline (Hoppe and Effland, 1998; USDA, 1999c).

The changes in the modern food and agriculture system pose major challenges for publicly funded research and education. One major challenge is in meeting the diverse and complex needs of agricultural producers—those in commercial agricultural production and those in the sector dominated by the multitude of smaller producers, including niche producers and low-income, limited-resource producers. Publicly funded research and technology are perceived to have favored increases in farm size and to have promoted industrialization of the farm sector. There is debate over whether publicly funded agricultural research is equally accessible to all users and whether it is targeted to the full range of user and citizens' groups.

This report analyzes the impacts of agricultural research on farm structure, and it offers recommendations related to research and extension programs. It evaluates the applicability of publicly funded agricultural research across the agricultural sector, which ranges from small, poorly capitalized farms to large, well-capitalized industrial organizations. The focus is on analysis without judgment about the social desirability of particular distributions. The committee worked under the assumption that one of the mandates of the public sector is to serve constituents, as demonstrated by the enabling legislation that established the public sector agricultural research system and, more recently, by a public policy that has increasingly supported constituent service. There is a legitimate concern that adoption of some of the committee's recommendations might result in reduced economic surplus in the aggregate. However, the committee submits that how losses or gains are distributed is also an important question. The issue of distribution among participants in the agricultural industry is a critical and largely missing component of the public policy debate concerning R&D and innovation investments.

This chapter provides background on the structural changes occurring in U.S. agriculture; an overview of the committee's charge and study process, a look at related literature; a brief description of the dimensions of farm structure; background on publicly funded agricultural research; and an organizational summary of this report.

THE STUDY PROCESS

The U.S. Department of Agriculture (USDA) Research, Education, and Economics Mission Area requested that the Board on Agriculture and Natural Resources of the National Research Council (NRC) convene a panel of experts to examine how publicly funded agricultural research has influenced the structure of U.S. agriculture. The panel was asked to assess the influence of public-sector agricultural research on changes in farm size and numbers, with particular emphasis on the evolution of very-large-scale operations. The committee's charge was as follows:

- Review relevant literature—including pertinent rural-development literature—on the role of research and the development of new technology in promoting structural change in farming, evaluating theoretical and empirical evidence.

- Consider whether public-sector research has influenced the size of farm operations and, if so, by what means.
- Provide recommendations for research and extension policies, giving consideration for access to results of public-sector research that leads to new farm production practices and technology.

The committee analyzed publicly funded agricultural research documented in the Current Research Information System (CRIS) database, which is USDA's documentation and reporting system for research projects in agriculture, food and nutrition, and forestry. It also considered information drawn from case studies and from the scientific literature. The committee gathered input and information from stakeholders during two public workshops held in conjunction with this study (Appendixes A and B).

This study complements several NRC reviews and evaluations of federal and publicly funded agricultural research programs: *National Research Initiative: A Vital Competitive Grants Program in Food, Fiber, and Natural-Resources Research* (2000b); *Sowing Seeds of Change: Informing Public Policy in the Economic Research Service of USDA* (1999); *Colleges of Agriculture at the Land Grant Universities: Public Service and Public Policy* (1996); *Colleges of Agriculture at the Land Grant Universities: A Profile* (1995); *Investing in the National Research Initiative: An Update of the Competitive Grants Program in the U.S. Department of Agriculture* (1994); and *Investing in Research: A Proposal to Strengthen the Agricultural, Food, and Environmental System* (1989). Those reports were produced, in part, to assess the adaptation of publicly funded research to changing needs and priorities and to enhance the ability of publicly supported agricultural research to serve the national interest.

This study also builds on the findings and recommendations of other reports that address structural issues in agriculture: *National Agricultural Research, Extension, Education, and Economics Advisory Board Recommendations* (2000a, 2000b); *A Time to Act: A Report of the U.S. National Commission on Small Farms* (USDA, 1998b); *A Time to Choose: Summary Report on the Structure of Agriculture* (USDA, 1981); *Technology, Public Policy, and the Changing Structure of American Agriculture* (OTA, 1986); *Charting the Course for the Cooperative Extension System Federal Agenda: The Working Group Report* (USDA, 1996); *The Relationship of Public Agricultural Research and Development to Selected Changes in the Farm Sector: A Report to the National Science Foundation* (Busch, et al., 1984); *Report of the Strategic Task Force on USDA Research Facilities* (USDA, 1999e); and *Vertical Coordination of Agriculture in Farming-Dependent Areas* (Tweeten and Flora, 2001).

STRUCTURAL CHANGES IN U.S. AGRICULTURE

The "structure of agriculture" is a broad phenomenon involving both the characteristics of farms[1] (the system of agricultural production) and the relationships of farms to other sectors and institutions. "Structural change" refers to the change in those characteristics over time. While there is no universally accepted definition, many authors have described the structure of agriculture (Boehlje, 1992; Breimeyer, 1991; Lee, 1980; USDA, 1981; Weber, 1978), and there is considerable agreement that farm structure involves matters such as:

- The size (measured in acreage or gross farm sales) and size distribution of agricultural operations, including the concentration of agricultural production—the increasing share of agricultural output by fewer and fewer firms.
- The number of agricultural operations.
- The spatial character of production systems.
- The technology and production characteristics of agricultural operations, including the level of specialization and diversification.
- Resource ownership arrangements, including tenancy and leasing.
- The relationship among ownership, management, and labor.
- Dependence on primary resources, or the relative degree to which an operation depends on capital, labor, or knowledge.
- Inter- and intrasectoral linkages, including contractual relationships for marketing and production.
- The extent and pattern of vertical and horizontal integration.
- Production, marketing, and financial management strategies.
- Business organization (including sole proprietorships, partnerships, corporations, and cooperatives) and arrangements (including joint ventures, leasing, independent production, and contracting).
- Characterization of the workforce, including age, gender, ethnicity, education, experience, skill level, and part-time versus full-time status of the operator.

In recent years, the dimensions of farm and agricultural structure that have been of greatest interest in public policy discussions have been farm numbers, the size distribution of farms (and concentration of farm production), and relationships of vertical integration among input providers, farm producers, and agricultural

---

[1] By joint agreement among the USDA, the Office of Management and Budget, and the Bureau of the Census, a farm is any place from which $1,000 or more of agricultural products (crops or livestock) was sold during the year under consideration. This definition has been in place since 1975 (Sommer et al., 1998).

INTRODUCTION

processors. This report will emphasize these structural characteristics but will also consider other dimensions of structure such as the spatial distribution of production, spatial specialization of production, and part-time farming.

The following terms are frequently used to describe farm structure:

Small farm: Until 1997, the Economic Research Service (ERS) used $50,000 in agricultural sales as the line of demarcation between large and small or "commercial" and "noncommercial" farms. The National Commission on Small Farms, in contrast, established by the Secretary of Agriculture in 1997, used $250,000 in gross sales as its cutoff. Under that definition, 9 out of 10 U.S. farms would be considered small. ERS has since incorporated the $250,000 cutoff in a new farm classification system that divides U.S. farms into mutually exclusive and more homogeneous groups (Hoppe et al., 2000). The ERS farm typology is presented in Appendix D.

Family farm: Although many debates about the structure of agriculture involve the future of the family farm, there is no generally accepted definition of the term. The ERS definition includes farms organized as proprietorships, partnerships, and family corporations, but it excludes farms with hired managers and farms organized as nonfamily corporations or cooperatives (Hoppe et al., 1996; Salant et al., 1986). By the ERS definition, a family farm is one in which the operator and the operator's household legally control the farm. On farms not included under the ERS definition, the farm manager and manager's family have limited authority over the distribution of the net income or the equity of the farms they operate. The U.S. Census of Agriculture defines family farms more broadly to include partnerships and family-held corporations, as well as sole proprietorships, with no restrictions on hired managers (USDA, 1999c).

Vertical coordination: Vertical coordination provides synchronization or coordination of two or more stages in the production and marketing chain under common ownership via management directive (Martinez, 1999). This implies more closely coordinated value chains and fewer alternatives for selling agricultural products.

Integrated Ownership: This is the major form of vertical integration, in which a company owns and operates, in addition to input supply or food processing and marketing, crop or livestock production in at least one stage of the food production chain (Tweeten and Flora, 2001).

Horizontal consolidation: Horizontal integration, or consolidation, is the process in which businesses that produce the same product at the same stage of the food marketing chain merge or are joined through acquisitions (Tweeten and Flora, 2001). Consolidation is measured in terms of farm size and number of processors, number of input suppliers, and number of retailers.

Regional distribution: The geographic distribution of agricultural production.

Industrialized production: This is large-scale production using standardized technology and management linked to processors by formal or informal arrangements.

Underserved farmer: The underserved farmer has limited access to the factors of production (land, labor, and capital). Social class, race, ethnicity, and gender are related to underserved farmers' decreased access.

PUBLICLY FUNDED AGRICULTURAL RESEARCH

Agricultural research—whether funded in whole or in part by federal, state, and local government sources—was established to serve the public good. Indeed, numerous legislative actions (beginning with the Morrill Act of 1862, the Hatch Act of 1887, and the Smith-Lever Act of 1914) have been taken to establish a system for research and outreach that would serve rural people and agricultural producers as a whole. Publicly funded agricultural research includes a complex variety of funding sources and users.

Sources of Public Research Funds

The committee considered *publicly funded agricultural research* any agricultural research performed with financial or material support from the public sector, including international organizations, federal agencies (e.g., USDA, the U.S. Geological Survey, the Department of Interior, the Department of Energy, or the Environmental Protection Agency), state governments and

---

BOX 1-1 Public Sources of Funding for Agricultural Research

Legislative commissions
Municipal airports, townships
National Aeronautics and Space Administration
National Cancer Institute
National Institutes of Health
Continues

> *Public Sources of Funding...continued*
> National Park Service
> National Renewable Energy Lab
> National Science Foundation
> Park and recreation boards
> Regional fishing, state departments of agriculture
> River basin commissions
> State boards of water and soil resources
> State departments of administration
> State departments of health
> State departments of natural resources
> State departments of public service
> State departments of transportation
> State pollution control agencies
> U.S. Army, Bureau of Land Management
> USDA Forest Service
> U.S. Department of Defense
> U.S. Department of Energy
> U.S. Department of the Interior
> U.S. Environmental Protection Agency
> U.S. Fish & Wildlife Service
> U.S. Geological Survey
> U.S. Peace Corps
>
> (Source: Survey of the University of Minnesota College of Agriculture, 2000)

agencies (e.g., state departments of natural resources, boards of water and soil resources), and local governments. Box 1-1 shows the variety of public funding sources for agricultural research surveyed by the committee at one college of agriculture. The proportion of public funds in any research activity varies by institution and project.

Institutions Performing Publicly Funded Agricultural Research

Many institutions conduct publicly funded agricultural research. However, the committee elected not to survey and analyze comprehensively all sources of publicly funded agricultural research, given the challenges in defining and disaggregating investments in agricultural research over time across different agencies and in determining their relationship to structural variables. The committee instead chose to limit the scope of its analysis to a subset of publicly funded agricultural research that could be used as a proxy for the wider scope of research described above. The committee chose to emphasize the principal components of USDA-supported agricultural research and extension, including extramural research by state-level partners and other programs administered by the Cooperative State Research, Education, and Extension Service (CSREES); intramural biophysical science research conducted by the Agricultural Research Service (ARS); intramural social science research conducted by ERS; the collection and analysis of agricultural data by the National Agricultural Statistics Service (NASS); and research conducted by USDA's Forest Service.

The committee considered a variety of extramural research, extension, and education programs supported by CSREES, states, and other partners. The State Agricultural Experiment Station (SAES)-land grant system, funded in combination by state, federal, and private sources, is the largest component of the U.S. public agricultural research system. Most public funding for SAES is appropriated by state legislatures. Although CSREES-administered formula funds, based on each state's share of total rural and farm populations, also support the SAES-land grant system, the USDA allocation of total SAES expenditures is small relative to that of the states. In FY 1999, state appropriations to SAES accounted for 51 percent of its total funds; USDA funded only 17 percent. Other federal institutions accounted for about 12 percent, and private sources made up the balance (USDA, 1999f).

PRIVATELY FUNDED AGRICULTURAL RESEARCH

Recent changes in the relative magnitude of public-sector and private-sector agricultural research provide important context for a discussion of structural change and publicly funded agricultural research. The committee acknowledges that private sources of funding for agricultural research have grown significantly relative to public sources (Huffman and Evenson, 1993; See table in Appendix C). Public-private partnerships are increasingly used as a mechanism for both research and technology transfer, through publicly administered, commodity-levied research and promotion, or "checkoff", programs, Cooperative Research and Development Agreements, (CRADAs), and other financing and institutional mechanisms (see Fuglie et al., 1996 for a review; discussed in greater detail in Chapter 3). The committee recognizes that mutual influence exists between private and public research, both through the input of public research results into private sector research and through the influence of the private sector on the public-sector research agenda through financing arrangements. The committee considers linkages between the public and the private sectors important and likely to have significant implications for the structure of agriculture. However, an analysis of the relationship between privately funded agricultural research and the structure of agriculture is beyond the scope of this report. It is a significant issue that should be the focal point of further analysis.

REPORT ORGANIZATION

The rest of this report summarizes the committee's analysis of the impact of agricultural research on farm structure.

Chapter 2 provides an analysis of the implications of research and technology for the structure of agriculture. The chapter first examines academic analyses of the impacts of research on structural change. Second, the chapter addresses the structural implications of specific types of research innovations, including mechanical, chemical, biologic, and managerial. Third, it describes structural changes associated with the Green Revolution; the introduction of the tomato harvester; and innovations in animal agriculture, including the use of recombinant bovine somatotropin. Finally, the chapter discusses the structural implications of research priorities, including criteria for setting priorities and for obtaining input from stakeholders.

Chapter 3 discusses the structural implications of technology adoption and technology transfer, including the influence of extension, the technology transfer arm of the public sector. This chapter also highlights new models for extension, including broader functions and new partnerships that should be further studied with regard to their structural outcomes.

Chapter 4 discusses the structural implications of public investments in agricultural research, using historic evidence and current data. The chapter highlights examples of public-sector responses to structural issues, including the development of new funding mechanisms and support for alternative agricultural technologies of interest to producers outside the commercial mainstream.

Chapter 5 frames public-sector research and development in the context of other drivers of structural change in agriculture, including market forces, public policy, and the changing role of knowledge and information. Chapter 5 documents changes in the agricultural sector that have resulted from these drivers. Finally, the chapter discusses the structural implications of drivers of change that should be considered in the design of future agricultural research and development policies.

2

Structural Impacts of Research

This chapter analyzes the structural implications of publicly funded research and technology development for U.S. agriculture. It first examines academic analyses. Then, the chapter addresses the structural implications of specific types of innovations, including mechanical, chemical, biologic, and managerial. Third, the chapter describes the structural impacts associated with combinations of innovations in three case studies that examine the Green Revolution, the introduction of the tomato harvester, and new developments in animal husbandry. Finally, this chapter discusses the structural implications of the process of setting priorities in agricultural research, including the criteria used and methods of obtaining stakeholder input.

RESEARCH AND THE STRUCTURE OF AGRICULTURE

Very little empirical evidence exists on the relationship between public sector agricultural research and structural change. Two studies examined the correlation between the extent or intensity of public research in the United States and the rate of increase in farm size and related structural characteristics. The first study, by Busch and colleagues (1984), examined the U.S. Censuses of Agriculture and other U.S. Department of Agriculture (USDA) data by state for 1915–1973. Indicators of the public research effort (including research expenditures and the number of research personnel) were related statistically to several indicators of farm size and to the concentration of agricultural production. A simultaneous-equation model was employed, using the state as the unit of analysis. The study provides strong statistical evidence that publicly financed agricultural research and development (R&D) is correlated with increases in average farm size, the number of very large farms (1,000+ acres), and large farms as a percentage of all farms (when controlling for a range of variables, such as farm mortgage debt, government payments to farmers, or value of marketed farm output). The largest effect is seen in the increase in the relationship between R&D expenditures and the percentage of large farms.

Another study, by Huffman and Evenson (2001), used a similar database and reached comparable findings. Huffman and Evenson used data from 1950 to 1982 from the Censuses of Agriculture, USDA, and related state-level sources. The authors used a six-equation econometric model and a large number of control variables to disaggregate the factors that lead to structural changes in farming. The objective was to estimate proportional differences over time in farm-structure-dependent variables attributable to three sets of variables: public R&D and education, private R&D and market forces, and farm commodity program payments. The indicators of structural change included crop and livestock specialization, an index of average farm size (essentially a normalized indicator of the average value of services obtained from physical capital and farmland), and amount of part-time farming. Huffman and Evenson reported that public research and education have been at least as important as private research and development and market forces for changing livestock specialization, farm size, and farmers' off-farm work participation over the study period. The strength of the relationship between public research and farm growth increased over the last third of the study period (from roughly the early 1970s to the early 1980s). Private R&D and market forces have been relatively more important than public research and education for changing crop specialization. Changes in farm commodity programs had little relationship to farm structure over the study period.

Although the Busch et al. (1984) and Huffman and Evenson (2001) studies represent different disciplines (rural sociology and agricultural economics, respectively), they have largely consistent results. In the aggregate, they associated the extent or intensity of the public agricultural R&D effort with an increased scale of agricultural production. However, these studies analyzed a limited set of variables (e.g., number, size, specialization, and farmers' off-farm work participation), whereas the study committee used a much broader definition of structure that includes a wide range of variables. Thus, it is difficult to make a general statement about the overall relationship between public-sector R&D and structural change based on these results. Furthermore, the Huffman and Evenson study demonstrates that the relationship of public R&D to different variables is mixed depending on the structural variable tested. They present evidence to support that public R&D is a major factor for some structural variables but not for others. It is thus difficult to assess the magnitude of the relationship between public R&D and structural change relative to that of other drivers of structural change.

INNOVATION AND THE STRUCTURE OF AGRICULTURE

The literature suggests that innovations that result from research vary in their influence on the structure of agriculture (Sunding and Zilberman, 2001; Thirtle and Ruttan, 1987). We can distinguish among mechanical, biologic, chemical, and managerial innovations, which also can be divided into those that increase yield or reduce costs of farming. The cost-reducing innovations can be subdivided into those that are labor saving and those that are capital saving. Related categories include innovations that augment human capital (automated management strategies) or that preserve natural resources. Modern irrigation technology, for example, can improve land quality and conserve water (Caswell, 1991). With the rise in consumerism, the importance of product-based innovations has grown, and there is much effort to improve the quality of food products. A related category of innovation is improved postharvest performance of agricultural systems, for example, that extend the shelf life of fruits and vegetables or that streamline shipping and handling. The environmental movement has raised the value of protecting environmental quality and of reducing the damage caused by agricultural activities. An additional category of environmental innovations detects damage and promotes better monitoring of farming and ecosystem performance (Millock et al., in press). Innovations in satellite imaging are expected to improve environmental decision making in agriculture (NRC, 1997b). This variety of innovations could have implications for the structure of agriculture. This section discusses innovations that influence horizontal consolidation, vertical integration, and regional distribution.

The concept of *scale neutrality* is used as a criterion for assessing the structural impact of a particular innovation. Scale neutrality is the ease with which a particular technology can be adopted, its divisibility into small enough units to be adopted, and its potential to benefit large and small producers alike in terms of results or relative profit. Scale neutrality is often confused with, but is distinct from, divisibility (see Chapter 3). Many divisible technologies, such as improved seed, which can be obtained in small or large quantities at the same unit price, are not necessarily scale neutral if they cannot be applied to a small-scale context.

Mechanical Innovations

Mechanical innovations typically are applied to farm machinery. Tweeten (1989) argued that mechanical innovations might have the greatest influence on horizontal consolidation, especially with regard to the increase of farm size and the reduction in the number of farms. Mechanical innovations contribute to horizontal consolidation for two reasons: First, they tend to reduce the requirement for labor—a main input provided by the farmer. Second, the capital cost per acre declines as the size of the farm increases. Capital-cost advantages favor large farmers for additional reasons, as well. Owners of larger farms often have easier access than do small operators to the capital they need to invest in equipment.

Mechanical innovations were important in our transformation from an agrarian to an urban society. Tractors and harvesting equipment saved labor, and they contributed to the increase in the size of farm operations (Kislev and Peterson, 1996). The trend continues, as new machinery dictates an increase in the size a farm must be to achieve economic viability—particularly in field-crop production.

There is a perception that large commercial farms are capital intensive and that smaller farm operations are more labor intensive. Although this is true in most cases, some of the large-farm operations in the country—Dudda Brothers in Florida and some major fruit and vegetable growers in California, for example—are very labor intensive. U.S. agriculture relies heavily on immigrant labor, particularly for harvesting, because the wage is too low to be appealing to many native-born American workers. Dependence on immigrant labor creates the problem of legal and illegal immigration. Automation of harvesting can actually lead to a reduction in manual, low-skill jobs and create higher paying jobs that attract nonimmigrant workers and, as is the case with the California lettuce industry, involvement with organized labor (Martin, 1985; Martin and Perloff, 1997).

Labor-saving machinery, such as the cotton harvester, has had other important social implications. Automation and mechanical innovations have been important for the viability of part-time farmers (Bessant, 2000; Huffman and Evenson, 2001; Kislev and Peterson, 1996). Parker and Zilberman (1996) have found that part-time farmers who own small orchards are among the first to adopt the use of computerized irrigation. Caswell et al. (1984) demonstrated that these types of operations are also among the first to adopt drip irrigation.

The operation of agricultural machinery, especially tractors, has been identified as among the most hazardous occupational activities in the United States and around the world (Forastieri, 1999). Some industries, such as the mining industry, tend to encourage the substitution of capital for labor to maximize worker safety. So innovations that increase safety and improve the well-being of farm workers can make the industry more attractive to potential employees.

Mechanical innovations are important in fostering environmentally sustainable forms of agriculture. For example, equipment is required for the transition from conventional tillage to reduced tillage systems. Some mechanical innovations, including those that use computers, can be important for monitoring the environmental impact of agricultural systems. Mechanical innovations also are being put to use for waste management.

Despite the importance of the mechanical sector in agriculture, the public research contribution in this area has not been substantial. Most mechanical innovations, many of them in farm machinery, have been introduced by the private sector (Feder et al., 1985). As we demonstrate in Chapter 4, only a small amount of public expenditure is devoted specifically to mechanical innovations for agriculture, although work supported by other public entities, including the military, the National Aeronautics and Space Administration, and the Department of Energy, could have spillover effects in agriculture.

Chemical Innovations

Chemical innovations, such as the development of herbicides that tend to replace labor, are likely to favor large farms, although that would mainly be the result of volume discounts for procurement. The volume used per acre would be the same for large and small farms. Application costs per acre likely decline with size, but that phenomenon is a consequence of mechanical application (Feder et al., 1985; Thirtle and Ruttan, 1987).

Examination of the structural effects of chemical innovations in agriculture shows mixed effects: Most chemical solutions are divisible, easy to apply, and usable even by the smallest farms. Mid-size or small family farms with field crops (corn, cotton) often are large enough to afford aerial spraying, and many chemicals are applied by certified applicators who charge on a size-neutral, per-unit basis. Some larger farmers receive volume discounts or own their equipment, so they save resources in application, but these have not been documented to be major advantages. The simplicity and low labor intensity of chemical treatments can benefit mid-size family farms operated by older farmers. On the other hand, the adjustment and education associated with pesticide regulation and the transition to new pesticide application techniques all require human capital and effort, which could favor the more dynamic commercial farms and lead to older farmers' exit from farming (Green, 1995; McWilliams and Zilberman, 1996; Putler and Zilberman, 1988). Finally, the use of chemicals, particularly herbicides, allows fewer operators to manage more acres.

Much of the basic chemical and engineering research that led to the use of pesticides, fertilizers, and chemicals occurred outside colleges of agriculture and agricultural experiment stations. Most recent pesticide research has been performed in the private sector. Annual sales of chemical pesticides in the United States were $12 billion in 1997, of which about 7 to 13 percent is estimated to have been spent on private research and development. Development, testing, and registration of a chemical can take 8 to 12 years and can cost over $50 million for each pesticide (NRC, 2000b). Land grant scientists and Agricultural Research Service (ARS) researchers do not emphasize basic research in chemistry. The public sector invests in pest management research but seldom in research on chemical toxicity and effectiveness. Public-sector work complements that of the private sector, and in some cases, the chemical industry may pay researchers or specialists to test new products.

Innovations in Biology

Innovations in biology, especially new seed varieties, increase general yield and can actually increase yield per acre. Their use requirements do not vary with the size of the farm. On those grounds, they have more neutral structural impacts than do mechanical or chemical innovations. On the other hand, some innovations in biology, for example, the new tomato varieties introduced to complement the tomato harvester, have significant structural effects in that larger farms might have an economic advantage in adopting new innovations because of fixed costs associated with education, capital requirement, and other factors (Feder et al., 1985). The literature on adoption suggests that when new crop variety properties differ significantly from those of traditional varieties, adoption can require drastic changes in the production system (Mann, 1978). For example, adoption of Green Revolution high-yielding varieties[1] often requires fertilizers and irrigation to be profitable and effective. Green Revolution varieties have been associated with high fixed costs for education and adjustment, which in the short term confer a competitive advantage to more affluent, better educated growers (Thirtle and Ruttan, 1987). Introduction of new varieties will have a smaller structural effect when the new varieties are closely related to old ones. Introduction of more drastically different varieties can have differential effects according to farm size.

New applications of biology to agriculture have resulted in seeds that obviate the use of chemical pesticides (*Bt* cotton[2]) or that augment their use (Roundup Ready varieties[3]). Innovations in biotechnology could permit gradual

---

[1] High-yielding varieties of wheat, rice, and maize adopted in the developing world during the 1960s and 1970s.
[2] Transgenic plants containing endotoxin genes from the bacterium *Bacillus thuringiensis* (*Bt*) conferring resistance to moths and butterflies (lepidopterans), flies and mosquitoes (dipterans), or beetles (coleopterans), depending on the class of *Bt* protein.
[3] Transgenic plants resistant to Roundup (glyphosate), a broad-spectrum, nonselective systemic herbicide.

modification of existing varieties and, thus, might not require significant adjustments or changes in the structure of the system. *Bt* corn and Roundup-resistant soybeans, for example, were adopted at high rates because adoption did not require significant adaptation of the production system. Pray (1993) suggested that the genetically modified varieties have higher rates of adoption and are accepted by smaller farmers because they are simple and convenient to use. Convenience, low cost, timesaving, and simplicity have been cited as reasons for the high adoption rates for genetically modified varieties in the United States (Carpenter and Gianessi, 1999; Fulton and Keyowski, 1999; Hubbel et al., 2000). However, more stringent environmental regulations associated with biotechnology would likely reduce adoption and produce fixed-cost and knowledge requirements that could deter some farmers. In addition, as chemicals, pharmaceuticals, and other exotic materials are increasingly produced using biotechnologic tools, vertical integration and contract farming are likely to result as the private sector buys the rights to and develops marketing strategies for these innovations (Zilberman et al., 1999).

Innovations in biology also could significantly affect the structure of agriculture because of differential benefits across regions. Olmstead and Rhode (1993) have demonstrated that innovations in biology have enabled the crop production system to adapt to different ecologic zones and have contributed to the growth of agricultural productivity in the United States. New heat- or cold-tolerant traits can permit the expansion of production of certain crops (Caswell et al., 1984). For example, drought-tolerant Hass avocadoes and cherry tomato varieties have permitted production of those crops in arid regions of California. Similarly, frost tolerance traits have expanded strawberry production. We could see a trend toward consolidation and vertical coordination in agriculture if new farm operations in these locations are larger or are part of agribusiness organizations, as was the case with the expansion of tolerant varieties of specialty fruits to major growers in the West. However, introduction of new varieties that benefit small local growers in some regions could have the opposite effect.

Increased product differentiation in agriculture and the introduction of new varieties, each with unique features and attributes, can contribute to increased value added and profitability of agriculture. Innovations in biology could be an important source of increased diversity among crop varieties. Increased differentiation among varieties will contribute to increased differentiation among agricultural products and could enable farmers to have a wider choice of agricultural inputs and ways to respond to variations in weather, soil, and pest conditions. However, introducing new differentiated varieties to the market is not easy and could require vertical coordination. Many farmers will be unwilling to adopt new varieties that result in a distinct and differentiated product unless there is a buyer for this product. A private company that owns the right to genetic materials is likely to contract with farmers to grow the product and then buy it from them to sell it down the marketing chain. Calgene, for example, contracted with growers for the cultivation of Flavr Savr tomatoes[4]. New differentiated varieties can increase value added to agriculture and are also associated with increased contract farming.

Public research historically has emphasized developing and improving genetic materials (Busch et al., 1995). Because capturing the benefits from new seeds was difficult for the private sector, genetic selection and variety improvements fell mainly to public-sector research. The role of the private sector has expanded over time, however, beginning with the introduction of hybrids in the 1930s. The introduction of plant breeders' protection legislation in the 1970s (U.S. Congress, 1970) and the Supreme Court decision in 1980 (Diamond v. Chakrabarty, 1980) that allowed patenting of life forms gave private companies the legal tools to protect their investments in developing genetic materials (Wright, 1998). Indeed, we have seen an emergence of seed industries that coexist with the public sector in producing genetic materials.

The public-sector research and development effort is linked closely to the Consultative Group of International Agriculture Research (CGIAR) centers, which house seed banks and other genetic materials and conduct exchange programs among breeders in different countries. Alston et al. (1995) documented the significant economic benefits the United States has obtained from investments in new varieties and strains developed in other countries. However, access to genetic diversity could become limited and more expensive with the privatization of genetic materials.

Managerial Innovations

Changes in prices, weather, technology, institutions, and personnel can influence resource allocation and profits in the agricultural sector. Effective managerial practices consider all of those factors. Research on many management practices is needed to prevent faulty resource allocations, reduce the public and private costs of errors that do occur, and increase efficient use of resources. Jensen (1977) reported that managerial research focuses on assisting decision makers to use resources efficiently; helping policy makers determine the consequences of alternative policies; studying the economic effects of technologic and institutional changes on agricultural production and resource use; and studying individual farm, area, and regional adjustments in resource use.

Managerial innovations help farmers run their operations. One recent example is precision farming, which uses Geographic Information System/Global Positioning System (GIS/GPS) technology to ascertain field characteristics and minimize the use of irrigation or fertilizer, for example. Because of the high initial investment in equipment, this technology has been

---

[4] A transgenic tomato line with altered fruit ripening.

affordable only for large farms. Savings in input use may offset the initial outlay, but there is no consensus on this. Some studies argue that precision farming results in increased yields, reduced input use, and reduced environmental damage from excessive chemical use (Kitchen et al., 1996; Koo and Williams, 1996; NRC, 1997b; Sawyer, 1994; Watkins et al., 1998), but other studies demonstrate mixed results of precision farming on profitability (Carr et al., 1991; Swinton and Lowenberg-Deboer, 1998). Profitability studies may not be conclusive until spatial econometrics, whole-farm analysis, and management information system analysis are included in economic analyses (Olson, 1995). The effect could be mitigated if equipment suppliers are willing to provide custom service or rental arrangements (NRC, 1997b).

Vertically integrated structures, particularly in animal production, provide another example of managerial innovation. The relatively new broiler production industry, for example, developed from contractual arrangements between the grain industry and poultry farmers in which safe markets were ensured for the broiler producers in exchange for their guarantees of feed purchase from the grain industry (McBride, 1997). Hog farming also exhibits a high degree of vertical integration (Martinez, 1999). Vertical integration demonstrates mixed structural impacts: It can benefit small producers by reducing overall business risk, controlling costs, gaining and improving market position, and facilitating access to information and financial resources necessary to develop new crops. However, contract farming creates its own risk, despite reducing others. This risk is associated with failure to produce to contract standards, loss of independence, and weak bargaining power in negotiating contracts. Therefore, farmers who are unwilling to enter contractual agreements may be forced out of business (Rehber, 1998).

Socioeconomic data from government sources can be used to enhance private and public managerial decisionmaking. USDA's Economic Research Service (ERS) and National Agricultural Statistics Service (NASS) regularly compile and publish data on prices, quantities supplied, and quantities consumed, for example. USDA's Agricultural Marketing Service (AMS) also provides more time-sensitive information on a daily basis through its Market News Program. Those data can be used by farmers in making decisions, especially about production. Although the data are publicly available and their use would seem to be scale neutral, the value of the information for various farmers depends on their ability to access and use the data. The factors that affect decisions to adopt technology discussed in more detail in Chapter 3.

Advances in computer and information technology have been dramatic over the past 15 years and have contributed significantly to reducing the costs of management throughout the global economy, including the agricultural sector (World Bank Group, 2000). Putler and Zilberman (1988) analyzed survey data from Tulare County, California to demonstrate that size of the farming operation, education level, age level, and ownership of a farm-related nonfarming business significantly influence the probability of computer ownership. The type of application software owned is influenced primarily by the type of farm products produced, the size of the farming operation, ownership of a farm-related business, and the education level of the farmer. Other studies have found similar relationships between education and age level of users and the adoption of computers and the number of software applications used (Batte et al., 1990; McWilliams and Zilberman, 1996).

It is difficult to judge the scale neutrality of managerial innovations. Some (precision agriculture) appear to benefit larger farms. Others (vertical integration) have led to large structural changes in animal production and show mixed effects on the production abilities of smaller farms. Other innovations (price information) appear scale neutral but depend on farmers' ability to adopt them. Thus, managerial innovations can cause significant structural changes, but not necessarily to the exclusive benefit of larger farms.

INNOVATIONS APPLIED

Structural impacts associated with several important applications of combinations of innovations can be described.

Green Revolution

The impacts of public research on farm size and structure were addressed in the debate over the social consequences of the Green Revolution (see Lipton and Longhurst, 1989, for comprehensive review).[5] The Green Revolution debate was initiated about South Asia (particularly India and Pakistan), an area in which there were pronounced inequities in agricultural resources and rural political power. The use of short-stature "miracle wheats" in South Asia from the mid-1960s through the early 1970s resulted in extraordinary yield increases, typically a two-fold or larger increase in output per hectare over a short period (2–5 years). Dramatic increases in yield and the potential for applying the technology to famine relief in South Asia led to wheat breeder Norman Borlaug's winning the Nobel Peace Prize in 1970—perhaps the all-time high-water mark of global public research in agriculture. A concurrent trend toward increased loss of land

---

[5] At least as far as the classic Green Revolution was concerned, the major centers of research were not public research institutions in the conventional sense. The research performers were the International Agricultural Research Centers (IARCs) of what has come to be designated as the Consultative Group on International Agricultural Research (CGIAR). IARCs are funded primarily by governments and by philanthropic foundations from the developed world. IARC research can be considered public (or quasi-public) because there has been little private funding. IARCs operate under basically the same public-domain principles as do nationally based public research organizations (e.g., ARS and the State Agricultural Experiment Station system), and they are considered to be not-for-profit enterprises.

and increases in farm size and concentration was observed in South Asia. The critics of the South Asian Green Revolution argued that landlessness and increased wheat output were causally related and that similar effects of the Green Revolution were beginning to occur in the other major Green Revolution crops and regions[6]. Similar effects were observed in Asian rice and in Latin American maize production (Pearse, 1980). Proponents of the Green Revolution argue that on the whole, benefits of reduced food prices and increased food security outweighed the adverse structural impacts (Ruttan, 2000). Heated debate over the structural implications of the Green Revolution continues to this day.

Tomato Harvester

At virtually the same time that concerns were first being raised about possible structural consequences of the Green Revolution (see the history in Lipton and Longhurst, 1989), Jim Hightower's *Hard Tomatoes, Hard Times* chronicled similar issues with regard to U.S. public research[7]. *Hard Tomatoes, Hard Times* was not a formal academic study, but it suggested that the role of the University of California in developing the mechanical tomato harvester, which would be substituted for farm worker labor, was an example of how public research could have structural effects. Most land grant administrators and many land grant scientists criticized Hightower's exposé of the land grant research and extension system, generally claiming that land-grant-developed technologies are usually scale neutral, and thus unbiased.

While for many observers, the Hightower book first raised the issue of the structural impacts of public research on tomato harvest mechanization, the still-classic study on the topic by Schmitz and Seckler (1970) had been published several years earlier. In addition to the Schmitz and Seckler study, there has been a large empirical literature on public research and the scale of agricultural production in the tomato sector (Berardi, 1984; DeJanvry et al., 1980; Friedland and Barton, 1975; Friedland et al., 1981). These studies show that in response to the threatened termination of the Bracero Program[8], which provided inexpensive Mexican labor to California tomato producers, University of California agricultural engineers assisted in bringing to market a mechanical harvester that largely mechanized the harvest of processing tomatoes. These studies show that

---

[6]Brokensha et al. (1980) argue that the Green Revolution also supplanted indigenous cultivars and knowledge systems.

[7]"Hard tomatoes" was a reference to the breeding of tomato varieties with thick, tough, fibrous skins that could survive mechanical harvesting and were integral to the success of tomato harvest mechanization. "Hard times" was a reference to the loss of jobs by farm workers and small- and medium-sized tomato producers.

[8]The Bracero Program provided inexpensive Mexican labor to California tomato producers for more than two decades after World War II under the 1942 Mexican Farm Labor Program Agreement between the United States and Mexico.

the development and adoption of the tomato harvester dramatically altered the structure of the processing tomato sector, resulting in declines in the numbers of farms[9] and increases in the average scale of production and in the concentration of this sector. By contrast, there was relative stability in the structure of the fresh tomato sector where the tomato harvester was not widely adopted.

There is little doubt that the University of California's role in the development of the tomato harvester contributed directly and decisively to the increased scale of tomato production in California (particularly in processing tomatoes). It is not clear that the public sector was fully responsible for increasing the scale of agriculture in this case, for several reasons. First, during the 1960s, and even today, land grant, State Agricultural Experiment Station (SAES), or ARS funding of farm mechanization research has been relatively small. Thus, the University of California funding of harvest mechanization equipment research was anomalous by national standards. Second, tomato harvest mechanization had some social benefits, such as reduced consumer prices (Schmitz and Seckler, 1970). Thus, even the most well-researched case studies of the impact of public research on farm scale and structural change do not lend themselves particularly well to answering a basic question that is part of the focal point of this report.

Animal Agriculture

There has been relatively more research on adoption of new technology and methods to produce cultivated crops than there has been concerning animal agriculture. However, the most significant structural changes in agriculture have occurred in the livestock sector. The broiler industry and the swine industry, for example, present a new mode of industrial agriculture characterized by contracting, vertical integration, high concentration of animals, and increasing returns to scale. The dairy sector has experienced significant changes; large dairies in southern California, New Mexico, and Texas (with several hundreds or even thousands of cows that rely on prepared feed) have become an increasingly dominant segment of the market.

Many technologic and institutional innovations that led to increased regionalization, concentration, and vertical coordination in livestock production originated in the private sector. Public-sector research contributed little to technologies such as automated milking machines or to the herringbone-milking parlor, which increased the size of dairy farms. Similarly, automated poultry feeding systems resulted from mechanical innovations developed in the private sector in the United States and abroad.

---

[9]Heinicke (1994) also documents declines in farm numbers following the development of the cotton harvester.

Public-sector research could have contributed significantly to other areas of industrialized animal production. Data from research on diet, genetics, metabolism, and digestion were used by commercial producers to design feed formulae and develop industrialized animal production systems. Similarly, new information about disease prevention and animal health control facilitated increased animal density. Publicly supported discoveries on the manipulation of light to increase the egg-laying productivity of hens were also important in providing the economic rationale for industrialized egg production. In addition, public research institutions have contributed technologic innovations that increased the industrialization of agriculture. At Iowa State University, boars were selected and tested for their performance in confinement on concrete floors. Leg weakness and hoof shape were modified through breeding to improve their suitability for concrete floors (Hargrove, 1973; Rothschild and Christian, 1988).

Recent debate and scholarly investigation of the effects of public-sector research on the structure of agriculture have arisen surrounding the use of biotechnology, including recombinant bovine somatotropin (rbST)[10] and genetically engineered crop varieties. Biotechnology issues include matters of scale neutrality and scale bias and the implications of large-scale research agreements such as that consummated in 1999 between Novartis (now Syngenta) and the University of California, Berkeley[11].

Examination of the structural impacts of biotechnology, particularly rbST, has yielded two major findings. First, the extent of the public and private contributions to those technologies is extremely difficult to disaggregate. In the case of rbST, other than the Cohen-Boyer research on cloning genetically engineered molecules in cells (Cohen et al., 1973), most R&D was done commercially by Genentech, Monsanto, Cyanamid/American Home Products, Elanco, and others. Nevertheless, although there was relatively little public funding, SAES and land grant university scientists and research facilities were pivotal in the development of the technology. Public researchers tested rbST on their own herds and allocated scientist effort to its development. Later, Cooperative Extension encouraged its adoption.

A second major finding is that although rbST technology is divisible into small enough units, in principle, to be used on a farm of any scale, the pattern of rbST adoption is highly correlated with herd size. Using a comprehensive data set on rbST adoption in the United States, a 1999 Wisconsin study reported that

---

[10] A metabolic modifier produced using DNA technology that alters various physiologic functions, including the efficiency of milk production in dairy cows.

[11] In 1998, an alliance to support basic agricultural genomics research was signed between Novartis Agricultural Discovery Institute (NADI), a subsidiary of Novartis, and the Regents of the University of California, Berkeley, committing $25 million over five years in research funding to the Department of Plant and Microbial Biology faculty in the College of Natural Resources. In return, the company will receive first rights to negotiate licenses on a portion of patentable discoveries made, although UC generally retains the patent. The alliance went into effect in 1999 (University of California, Berkeley, 1998).

more than 70 percent of operators with herds of 200 or more cows used rbST (Buttel et al., 2000; Ostrom and Buttel, 1999). However, only about 4 percent of operators with herds of fewer than 50 cows used rbST. One critical feature of rbST use is that its benefits can be maximized only if high-quality feed is available and animal nutrition is managed accurately. Adoption of rbST is highly correlated with the use of other productivity-augmenting technologies (e.g., total mixed ration [TMR] equipment, which blends all feedstuffs [forage, grain, and supplements] to provide a complete source of nutrients in a ration) found almost exclusively on large dairy farms (Buttel et al., 2000). Thus, somewhat similar to Green Revolution crops, rbST is a highly divisible but scale-biased practice in terms of adoption rates across farms of different scales of production.

STRUCTURAL IMPLICATIONS OF THE RESEARCH PRIORITY-SETTING PROCESS

Criteria for Setting Priorities in Agricultural Research

The criteria used for setting priorities in public-sector agricultural research and the assessment of the payoff or benefits in the form of return on investment can have significant structural effects. Many of the criteria and payoff assessments have focused on productivity and efficiency goals, justified on the assumption that in itself, increased productivity will benefit producers and consumers and will feed the expanding global population. We now know that the issue is more complex. Hunger has many causes, including inadequate income to purchase food and distribution systems that are inadequate to transport food from producer to consumer. Achieving a goal of simply producing more food masks other problems related to expanding populations, including increases in disease and the destruction of natural resources vital to a productive and resilient ecosystem. Increased productivity in a market characterized by inelastic demand (where prices respond dramatically to small changes in quantities supplied) significantly reduces income for producers even when output increases, often reducing producer income to unacceptable levels.

The tension between serving diverse constituencies while promoting rapid, and often socially dislocating, increases in productivity is still at issue in public agricultural research well over a century after the public agricultural research and outreach system was established. The goal of publicly funded agricultural research and outreach in the United States as originally articulated in three acts of Congress—the Morrill Act of 1862 (U.S. Congress, 1862), the Hatch Act of 1887 (U.S. Congress, 1887), and the Smith-Lever Act of 1914 (U.S. Congress, 1914)—was to serve rural people as a whole. Together, these acts promoted:

"[T]he liberal and practical education of the industrial classes" (Morrill)

"[A] sound and prosperous agricultural and rural life," ensuring "agriculture a position in research equal to that of industry, which will aid in maintaining an equitable balance between agriculture and other segments of the economy" (Hatch); and

"[The] diffusing among people...[of] useful and practical information on subjects relating to agriculture and home economics, and [encouraging] the application of the same." (Smith-Lever)

Congress intended to use public agricultural research and outreach to help farmers and the "mechanic classes" to advance. Thus, social goals and service to a broad constituency were emphasized relative to goals of productivity or efficiency. These goals were an attempt to respond to the generally limited farmer support for research in the late nineteenth century. Farm journalists, chemists, and university administrators supported passage of research legislation (Danbom, 1986). In contrast, American farmers were more interested in seeking relief from their economic troubles (by curbing railroad monopolies, obtaining credit, expanding exports) than they were in supporting research or new technology (Marcus, 1985). In particular, the Hatch Act, which explicitly directed funds to be used for applied research, represented a compromise between the aims of its proponents (to modernize agriculture, "professionalize" the farmer, and harness science as an engine of national development) and the concerns of its opponents (to be practical and of broad benefit to rural people; Marcus, 1985).

It is important to acknowledge those who benefit in the long term from productivity and efficiency research in a commodity industry such as agriculture. Although the short-term benefits of advances in technology that improve productivity and efficiency can accrue to producers and other participants in the production and distribution chain, in a private-sector commodity industry characterized increasingly by global competition, the benefits are eventually captured by consumers, if no monopoly conditions exist. Although the focus of productivity-enhancing R&D might be on producers, the rest of the value chain, including consumers of food products, captures the payoff over the long term. Furthermore, the producer segment that captures the most benefits in a commodity industry is the one with the lowest expenses or the greatest control. If an industry is characterized by decreasing cost (as appears to be the case in agricultural production), then larger scale producers will naturally capture more of the benefits of productivity-focused research. We do not suggest here that productivity-focused research is inappropriate. We propose only that in a commodity industry characterized by decreasing cost and intense competition, the economic forces inherent in the market will in the long run transfer the benefits of productivity- and efficiency-increasing research to other parts of the value chain. Benefits retained will be those by the lowest cost, largest scale producers.

A second structural implication of productivity–efficiency criteria for funding and evaluating research is the limitation of these criteria to consider broader social goals in assessing the benefits of agricultural research. Those goals might include increasing diversity in agricultural production and distribution systems; reducing environmental and resource degradation; contributing to the long-term sustainability of agricultural production systems; improving the social well-being of producers and rural residents; and reducing financial, economic, political, and environmental risk. Only the public sector addresses these goals; their benefits cannot be captured by the private sector. Increasingly, this broader set of criteria is part of the allocation system for public-sector agricultural R&D expenditures. However, the fundamental distribution issue of who gains, who loses, and how the losses might be mitigated or repaid has never been a criterion for evaluation of public-sector research projects. Similarly, distributional consequences of research are not the focal point of the agenda for evaluating specific advances or innovations in agriculture.

As a larger proportion of the R&D budget moves into the private sector, the total public- and private-sector R&D budget in agriculture will focus increasingly on innovation that ignores the broader set of social and public goals but benefits consumers and efficient, large-scale producers. This will occur to the extent that productivity and efficiency criteria dominate the allocation and assessment process and as public-sector funding declines in relative proportion to private-sector funding of R&D. Private-sector R&D will be evaluated almost exclusively for productivity and efficiency, because that is how value is created and captured, and managers of publicly held companies are evaluated by generation of profits and the market value of stock. It should be acknowledged that, as segments of agriculture move from a commodity orientation to a differentiated-product orientation, lowest cost is no longer the only or dominant determinant of competitive advantage, and innovation that contributes to differentiation (rather than to cost reductions) has the potential to capture value. If producers of all sizes can adopt such an innovation, it can be less size biased than a productivity–efficiency innovation in a decreasing-cost, competitive industry. Whether there will be size advantages to producing differentiated versus commodity products and who will capture the benefits from innovation in differentiated product markets remain unanswered questions. Furthermore, as long as most of the differentiation in the food production and distribution chain occurs beyond the farm gate, the benefits from innovation in differentiated products will be captured by those who produce that differentiation and not by farmers or producers.

A third structural implication of the productivity–efficiency criteria for research funding and evaluation is the inadequacy of narrowly defined criteria to measure total resource productivity and efficiency. Specifically, the criteria are typically measured and defined in terms of private cost. Because of important differences between private and public cost, including such externalities as long-term resource degradation and the social cost of human resource adjustments, the private-cost-driven, productivity–efficiency criteria will be biased against R&D that might reduce total resource productivity and efficiency but that increase long-term sustainability. Ignoring public cost as part of the criteria for R&D allocations and assessment will bias the allocation process toward innovations that reduce private cost at the expense of those that might reduce both public and private cost. In essence, even if efficiency and productivity are the only social goals to be achieved in public-sector R&D allocations, ignoring externalities and public cost will have important distributional and structural implications.

The committee encourages the public sector to develop broader criteria for evaluating and funding agricultural research that will help producers—particularly those producers outside mainstream agriculture who are unable to compete in commodity markets—obtain and retain market value. Those goals might include increasing diversity in agricultural production and distribution systems; reducing environmental and resource degradation; contributing to the long-term sustainability of agricultural production systems (see Pretty, 1995 for a more detailed discussion of sustainable agriculture); improving the social well-being of producers and rural residents; and reducing financial, economic, political and environmental risk. The committee recognizes that there are limits to the degree to which developing technology for "niches" is sustainable, since increased research and development on a niche product will increase the size of the market, invite entry by other producers, and thereby turn the niche product into a commodity product. The quest for higher value niche production technology and products is thus a perpetual one. Nevertheless, the committee notes that the relative contribution of public research in developing technologies relevant to small-scale farmers, organic farmers, and others outside the commercial mainstream is an important determinant of its support to those constituencies. Structural concerns should be better balanced with all other factors involved in setting the research agenda. Only the public sector can address these goals; their benefits cannot be captured by the private sector.

Recommendation 1
The goals of public-sector research should continue to be broadened beyond productivity and efficiency. Federal and state research should improve technology and information systems that benefit farmers in diverse production systems and circumstances, including part-time farmers, small-scale farmers, organic farmers, and value-added producers. However, limiting public-sector research to scale-neutral technologies is not sufficient to meet the needs of a diverse producer constituency. The public sector increasingly should assess the opportunities for R&D and technology transfer for those who are not served by the private sector.

The study committee cites the need for interdisciplinary work that will integrate biophysical sciences, social sciences, and the humanities as an avenue to achieving broad research goals. Farming systems research and extension, which have been primarily implemented in developing countries, illustrate how the integration of social science approaches has broadened the research agenda. In Colombia, collaboration between rural sociologists and bean breeders changed the focus of a bean breeding program from early-maturing to fast-cooking characteristics, based on needs-assessment research conducted with end users (Feldstein and Poats, 1989). The resulting bean varieties were better suited to the needs of small farmers, who were primarily women. In Peru, anthropologists working with agricultural engineers shifted the potato storage research program from complete-dark, off-farm options to partial-light, on-farm storage options. The resulting technologies reduced post-harvest losses and storage costs. More small farmers were able to retain potatoes longer after harvest until prices were high, resulting in increased incomes (Rhoades and Booth, 1982). In the United States context, social science research involving the end users of soil conservation technology also contributed to more effective interactions among strawberry farmers and agricultural engineers in extension and in the Natural Resources Conservation Service (Mountjoy, 2001).

More social science research must be integrated into priority setting. The committee envisions an important role for social science research on agribusiness and entrepreneurial enterprises other than farm management; the costs, benefits, and consequences of technology, including social, human, and community factors; and rural development, including lifestyles and opportunities for individuals and communities. Social science research on farm structure and production systems also can serve as a needs-assessment baseline in research decision making.

It should be noted, however, that many public-sector research institutions lack a significant social science research effort. For example, in 2001, ARS reported only 1 economist, 3 home economists, and no sociologists in its workforce of 1,980 research scientists (USDA, 2001d). Among a total workforce of 4,278 employees in USDA's Research, Education, and Economics Mission Area, 307 economists, 6 home economists, 29 social scientists, 8 sociologists, and 7 education professionals are reported (USDA, 2001b). Some international agricultural research centers, particularly the International Potato Center, the International Center for Maize and Wheat Improvement, and the International Center for Tropical Agriculture, have successfully and fruitfully involved social scientists in setting research priorities (Ashby and Sperling, 1995). Increased social science research would be a fundamental—perhaps even groundbreaking—public policy response in the United States.

Recommendation 2
The public sector should use an interdisciplinary approach integrating biophysical, social science, and humanities perspectives to determine structural consequences of research and to assess the research needs of a diverse clientele. The public sector, particularly ARS, should strengthen social science expertise in the areas of setting research priorities and assessing the distributional implications of research and new technology.

Stakeholder Participation

The study committee finds that public engagement in research supported with public funds is highly desirable, not only because it will make such work more accountable to the public, but because it has the potential to improve the public sector's ability to serve a diverse constituency. In the 1940 *Yearbook of Agriculture*, T. Swann Harding wrote that research conducted inside a professional science culture becomes "celibate" through isolation from the realities of life (Harding, 1940). The effective use of publicly funded research and monitoring of its effects require civic engagement. Research made more accountable to the public extends peer communities, giving scientists the "opportunity to test their work against a wider public and a wider variety of knowledge" (Raffensperger et al., 1999).

Participatory methods developed in the 1980s have been demonstrated to be valuable tools in understanding local people's needs and priorities with respect to agricultural innovation and technology development (Chambers, 1983; Pretty, 1995). These methods have been used successfully in the CGIAR and in other international contexts to determine the research agenda for plant breeding, crop, and natural resource management (CGIAR, 1999; CGIAR, 2000; WARDA, 2000), and research results from that agenda have benefited small farmers, women, and other underserved groups. Preliminary assessment of watershed management programs indicates that the most successful environmental outcomes have emerged from programs that had facilitated the involvement of local land managers in diagnostic appraisal, planning, implementation, and performance monitoring and evaluation (Thompson and Guijt, 1999). Very little research, however, is available on the structural outcomes of research conducted using participatory methods.

In the United States, a recent survey of agricultural research decision making reveals that stakeholder[12] involvement can work effectively (Dyer et al., 1999). Many states have had or are implementing opportunities for stakeholder participation, and federal public-sector agricultural research is starting to engage stakeholders in the agenda-setting process as a result of the legislative mandate in the 1998 Agricultural Research, Extension, and Education Reform Act (U.S.

---

[12] An individual or group who has interest in a particular process.

Congress, 1998). The 22 ARS national programs and CSREES have hosted stakeholders at public workshops for the past 3 years. The National Agricultural Research, Extension, Education, and Economics Advisory Board has engaged stakeholder participation at public hearings around the country.

The committee identified major challenges to participation from conversations with program staff and individuals who had attended its public workshops. There is a lack of coordination in methodology for engaging participation and interpreting stakeholder input in decision making across programs and agencies; it is difficult to obtain representation from regions, sectors, individual farmers, small farmers, commodity producers, and minority institutions; and it is a problem to secure travel funding for individual growers who cannot afford to attend meetings. The committee also found it difficult to obtain information on the Internet about the findings of the stakeholder participation sessions—other than a record of the date, location, and subject area. There was little information to explain how interested parties might become involved in future public workshops. The committee suggests that an examination of public participation strategies could be used to develop more effective approaches at the federal, state, and local levels.

Recommendation 3
To improve accountability to constituents, the public sector, at both the federal and the state levels, should continue to incorporate the knowledge and needs of stakeholders through genuine public participation in setting priorities for research and in implementing research projects; encourage broad-based participation on research and extension advisory boards to assess the relevance and importance of proposed research and extension programs and to ensure that priority setting is responsive to a variety of needs, particularly those that cannot be met by the private sector; conduct critical analysis and assessment of the methods used for engaging, interpreting, and incorporating stakeholder input into decision making; and take action to make the participation process more understandable and transparent to the public.

Structural Impact Assessments

Data and research on the relationship between public research and structural change are limited. *Ex ante* impact assessment research on prospective technologic thrusts and *ex post* research on recently commercialized technologies is most urgent when these technologies are likely to have major impacts on the structure of agriculture, the environment, food safety, or the relations between agriculture and consumers.

Recommendation 4
Public-sector research institutions, at both the federal and state level, should develop expertise and research programs devoted to analyzing the distributional implications and impacts of agricultural R&D for various groups of producers, using both ex ante and ex post research designs. The study committee endorses the public sector's earlier efforts in this regard and encourages continued development of this research base.

SUMMARY

In this chapter we discussed the structural implications of agricultural research. Empirical evidence suggests that publicly funded agricultural research and development correlates with increases in average farm size, the number of farms, the proportion of large farms as a percentage of all farms, livestock specialization, and off-farm work participation. We also discussed the structural impacts of various types of agricultural innovations and noted the degree of involvement of public research in their development. Using impact on labor and cost per acre as a function of size as criteria for measuring the effects of research on farm size suggests that mechanical innovations and, to a large extent, chemical innovations have more significant effects on size distribution. Innovations in biology are more divisible but can require high fixed costs associated with learning and capital expenses. Managerial innovations appear to have mixed structural impacts. Each category of innovation has varied consequences for structure. It is difficult to measure the exact contribution of public research in innovation because of the close interaction of public and private research and because research often begins in public institutions but is finished and brought to market by the commercial sector.

The discussion of case studies highlighted other issues that link publicly funded research and structural change, including the scale neutrality of research innovations (that is, the ease of adoption of a particular technology, its divisibility, and its potential to benefit large and small producers), the contribution of public research to vertical integration in livestock production, and the contribution of public research to shifts in regional boundaries for modern production agriculture.

A discussion of the structural implications of the research-priority-setting process described three structural implications of using productivity and efficiency criteria for research funding and evaluation. First, in a commodity industry, the benefits of research that increases productivity and efficiency will accrue to lowest cost producers. Second, productivity and efficiency goals fail to consider broad social goals in assessing the benefits of agricultural research. Third, productivity and efficiency criteria fail to adequately assess total resource productivity and efficiency.

Finally, the importance of an interdisciplinary approach and stakeholder participation in the priority-setting process was discussed as an avenue for serving diverse constituencies, and a proposal for research programs to assess distributional implications of agricultural R&D was offered.

3

Structural Implications of Technology Transfer and Adoption

The influence of an innovation on the structure of agriculture depends not only on the nature of the innovation but also on who will use it. The characteristics of producers or farm operations influence the degree to which innovations are adopted. The first part of this chapter examines how the heterogeneity of producers and farm operations influences adoption of technology and innovation, and one approach for responding to this heterogeneity is offered. The second part of this chapter examines the way in which extension, the public-sector arm for transferring agricultural research results, can influence adoption—and hence structural change—through its public education and information programs. This section presents evidence demonstrating that *what* is transferred, to *whom*, and *how* it is transferred can have significant distributional effects. Finally, this chapter presents evidence that, at the state and the local level, extension is increasingly acknowledging the importance of and attempting to serve a greater diversity of farmers and other end-users. We highlight here structural and functional characteristics, innovative processes, and collaborative models of a more broadly "engaged" extension that should be investigated with regard to their structural implications.

FACTORS THAT AFFECT TECHNOLOGY ADOPTION

The literature on the adoption of new technology in agriculture distinguishes various types of technologies and recognizes that heterogeneity among producers and farm operations affects what is adopted, to what extent, and when. This section discusses the barriers to adoption that can affect producers differentially.

Farm Size

Differences in farm size may influence technology adoption. Figure 3-1, for example, shows the extent to which several technologies were adopted by dairy farms of different sizes. Some innovations, such as management-intensive rotational grazing—intensively grazing a portion of a pasture followed by a rest period to allow the forage to regrow—are used more by smaller farms than by larger farms. Larger farms tended to adopt others, such as total mixed-ration equipment and the use of milking parlors.

One characteristic that can affect the degree to which farms of different sizes adopt various new techniques or technologies is *divisibility*. The literature distinguishes between *bulky* and *divisible* innovations. Bulky innovations—such as tractors, combines, and other farm machinery—require a significant initial investment but reduce variable cost. It makes economic sense for a given farm to purchase a bulky technology only if its scale is above a critical level. There is a general assumption—and there is some supporting evidence (Feder et al., 1985; Marra and Carlson, 1990)—that larger farms tend to buy and adopt bulky innovations early. Small farms also might adopt the innovation if they collaborate and purchase equipment through a cooperative, for example, or if they rent equipment from dealers or obtain custom service from contractors. Homesteaders in the early days of U.S. agriculture demonstrated that smaller farms could benefit from machinery rental and custom services (Cochrane, 1979; Gross et al., 1996). Today, there is widespread use of custom services for harvesting and land preparation (for example, leveling fields using lasers).

Some of the scale effects of technologies can be offset by institutional arrangements. Nevertheless, the introduction of bulky innovations affects the structure of agriculture significantly. The per-unit cost for the equipment owner is generally lower than for the renter or for the user of custom service. Those who purchase farm machinery also could have an extra incentive to augment the size of their operations to make full use of new equipment.

Many agricultural innovations are *divisible* in that they can be divided into small enough units to, in principle, be used on any size operation: chemical innovations (fertilizers, pesticides); biologic innovations (seeds, biologic pest controls); and managerial innovations (new techniques of pruning, modification of timing for some activities). Divisible innovations are ostensibly more scale neutral than are bulky innovations; indeed, in many cases, per-unit gain from the adoption of innovations, such as seed varieties, does not vary with size. However, adoption of divisible innovations can entail a large initial investment,

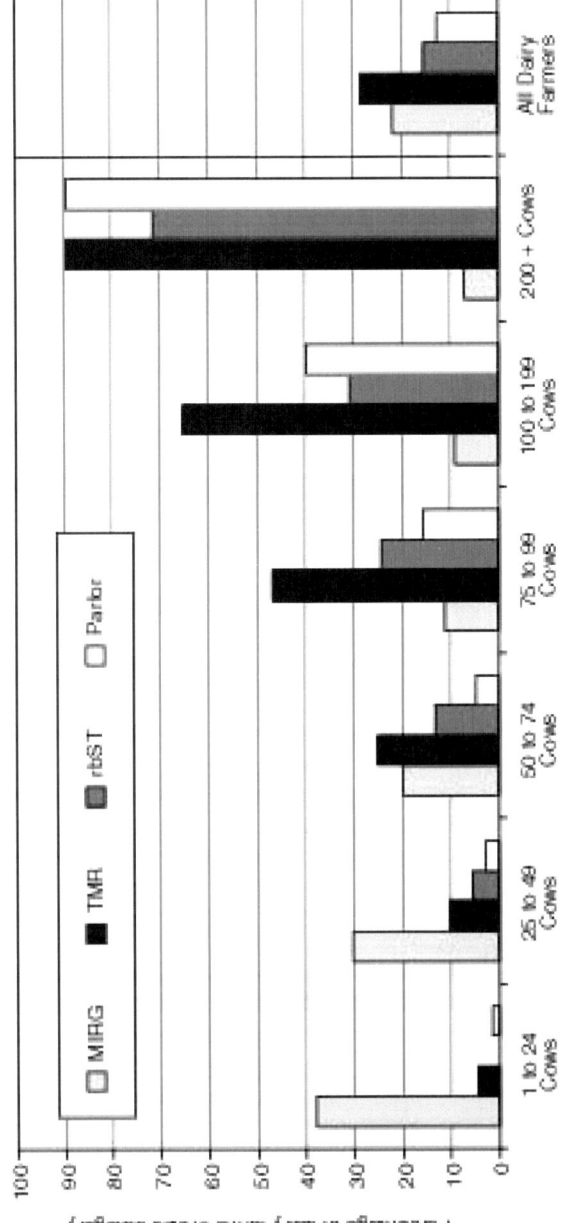

FIGURE 3-1 Technology adoption by dairy herd size, 1999. (MIRG, management-intensive rotational grazing; TMR, total-mixed ration equipment; rbST, recombinant bovine somatotropin; Parlor, milking parlor.)
SOURCE: Adapted from F. H. Buttel, D. B. Jackson-Smith, and S. Moon, 2000. A profile of Wisconsin's dairy industry, 1999.

putting some farmers at a disadvantage (Feder and O'Mara, 1981). The fixed cost can come in several forms. Adoption of some new technologies requires training, so farmers with small farms and farmers who are less educated could be at a disadvantage. Evaluation of new technologies is time consuming and, again, less-educated farmers or individuals with smaller farms could be at a disadvantage. Training farmers can require additional expense. For example, adoption of modern irrigation technologies, such as drip irrigation, can require fixed cost to redesign and modify farm operations, preventing some farmers from adopting the technology.

Just and Zilberman (1988) assessed the distributional effects of introducing new divisible technologies within farms of varying sizes. They separated farmers into four groups with respect to adoption of more profitable but riskier divisible technologies: farmers with the smallest farms who are unable to adopt because they cannot cover the fixed cost of learning and adoption; farmers with small farms who are limited in their capacity to adopt because of credit constraints; owners of mid-sized farms who can fully or almost fully adopt; and owners of large farms who could be partial adopters because of risk considerations. Smaller, nonadopting farms can be worse off in absolute terms if research and development that introduces new technologies leads to reductions in output price. The credit-constrained farmers also can suffer relative to mid-size full adopters. Just and Zilberman's analysis suggested that those who gain the most from the introduction of divisible technologies are farms large enough to fully adopt the technology but not so large that adoption will be hampered by risk considerations.

Regional Differences in Land Quality

The literature emphasizes the effect of differences in land quality and weather on technology adoption (Caswell and Zilberman, 1986; Green, 1995). Drip irrigation expanded California grape and avocado production to the foothills of the central and southern coasts and to sandy soils in Kern County. Center-pivot irrigation significantly expanded corn acreage to the sandy soils of western Nebraska and hillsides in Washington (Lichtenberg, 1989).

The development of such technologies that can benefit new farm owners in low-quality lands might not benefit owners of higher quality lands who do not need to adopt the technologies. In many cases, large farms on marginal land benefit from land-quality-augmenting technology (e.g., regions in the Central Valley in California), whereas small farms on high quality land, such as avocado growers in San Diego County in California, would not profit from the technology. The farms on higher quality lands are often well-established, traditional family farms that have been able to earn higher yields using less-advanced irrigation technologies. Thus, the research resulting in land-quality-augmenting technology has a significant distributional effect among different farms.

Human Capital

Human capital is another source of heterogeneity that has a significant influence on adoption in the context of rapid economic and technical change. Schultz (1975) describes two dimensions of human capital—*working ability* and *allocative ability*. Allocative ability is education level, intellectual skills, and aptitude for learning and assessing new technologies; working ability is physical capacity for labor. Many technologies are management incentives that draw on a farmer's allocative abilities. Huffman (1974; 2000) has related farmer schooling to decision making and adoption of technology. Wozniak (1993) has demonstrated that managers with more education are more likely to adopt new inputs and contact the extension service for adoption information than are operators with less education. Integrated pest management (IPM), for example, involves designing context-specific pest treatment as opposed to following a prescribed regimen of chemical pesticide application. Weibers (1992) shows that highly skilled farmers are more likely to adopt IPM, and, even after they seek the advice of consultants, educated farmers are likely to spray less pesticide and use the system more effectively.

Producer Age

Sunding and Zilberman (2001) reported that the tendency to adopt modern technology declines with age. McWilliams and Zilberman (1996) found that older growers owned fewer computers. Analysis of data from Tulare County, California showed that operator age level, along with education level and size of the farming operation, significantly influenced the probability of computer ownership (Putler and Zilberman, 1988). Green (1995) also found generational differences in modernization of farming practices, such as drip irrigation and automated irrigation, in California. Older farmers operate with shorter time horizons, so investing time and effort in adopting new innovations might not be practical. Younger farmers who operate with longer planning horizons often make a greater effort to acquire the skills or knowledge they need to adopt new technology. Older farmers might have less education or more limited familiarity with computers than do their younger counterparts. Older farmers, who own and operate a large percentage of farms in the U.S., are unable to take advantage of new technologies that are adopted by younger and more active farmers.

Tenure Arrangements

The literature emphasizes that contractual relationships, particularly tenure agreements (e.g., land ownership, rental), have a profound impact on adoption. Much of the research relating tenure arrangements to adoption has been performed in developing countries (Feder et al., 1985). Tenants with short-term contracts are less likely to adopt technologies requiring investment in land and assets that have a longer payoff. On the other hand, tenants might adopt equipment that is not tied to the land. Although farm management companies might not own land, they might own specialized, labor-saving equipment that will help them manage their operations more efficiently. Feder et al. (1985) surveyed a significant body of domestic and international literature and demonstrated that, under traditional tenure arrangements, when landlords are disengaged from farming activities and the contracts are of short duration, the adoption of modern technology is below average.

RESPONDING TO A BROAD VARIETY OF PRODUCERS, INCLUDING UNDERSERVED POPULATIONS

The previous section described factors that affect the adoption of innovation. These factors and others that characterize underserved groups, including race or ethnicity, are intangible policy parameters for research decision making. For example, there is such a diversity of small and medium-sized farms that it is difficult to generalize the essence of what they share in terms of research needs. It is much easier for a public research system to respond to the needs of underserved populations if it can target concrete production systems that have promise and can be funded readily. For example, in Wisconsin, management-intensive rotational grazing was used by about 22 percent of dairy farms in 1999, most of them small and medium-sized producers with herds of fewer than 100 cows (Buttel et al., 2000; Ostrom and Jackson-Smith, 2000). Hoop structures also have been used as a low-cost, labor-intensive alternative to confinement housing for smaller hog farms (Brumm et al., 1999). Thus, conducting research on rotational grazing or hoop structures might be useful for operators of small to medium-sized farms. If coupled with rigorous needs-assessment techniques to identify the production systems used by underserved populations, targeting nonmainstream or niche types of production systems can sometimes serve as a proxy for reaching underserved populations.

A public policy approach would engage public research organizations and administrators in developing packages of research and development activities that target specific clienteles (including groups defined by scale characteristics and scale-related characteristics, such as race or ethnicity). Within that framework, research administrators would be justified in allocating funds and resources to serve larger operators, provided the overall package of activities is transparent to the public and is balanced in terms of serving both medium-scale (or typical or representative) farmers and smaller, limited-resource or minority farmers.

Public-sector outreach activities, including extension, should serve a variety of producers, including limited-resource producers, organic producers, direct marketing producers, transition farmers, full- and part-time farmers, and cooperatives. The public sector should continue special efforts to reach out to underserved or minority communities.

Recommendation 5
Public-sector outreach, including extension, should take a proactive role in assessing the research and development and technology transfer needs of a variety of producers, including underserved and minority groups; designing appropriate strategies, such as applied, on-farm research, for serving those constituencies; and providing production assistance and other appropriate services, such as market development education for differentiated product markets, entrepreneurship education, financial strategies, value-added processing, and identification of opportunities for those working part time in agriculture.

The committee acknowledges that there are publicly funded research, extension, and education programs and projects for specific underserved populations, among them the U.S. Department of Agriculture (USDA) Small Farm Program, the Hispanic-Serving Institutions Educational Grants Program, the Tribal Colleges Research Program, and the 1890 Institutional Capacity Building Program. Minority-serving institutions, including the historically African American 1890s land grant universities[1], the 1994 Native American land-grant universities[2], and the Hispanic-serving institutions[3] have provided important access to underserved groups. Other institutions also are responding in innovative ways (see Box 3-1). However, the committee also submits that the public-sector response to these populations has been less than proactive, initiated

---

[1]The 1890s institutions were created as a result of the Second Morrill Act of 1890, expanding the 1862 system of land grant universities to include African American institutions. Today, there are 17 1890s institutions—including one private institution, Tuskegee University—located primarily in the Southeast.

[2]Tribal colleges were created over the last 30 years in response to the inadequate higher education rates of Native Americans and generally serve geographically isolated populations that have no other means of accessing higher education beyond the high school level. In 1994, Congress conferred land-grant status to thirty tribal colleges through the Equity in Educational Land Grant States Act of 1994 (U.S. Congress, 1994).

[3] Section 316 of the Higher Education Act Reauthorization (U.S. Congress, 1992) identifies Hispanic-serving institutions as accredited and degree-granting public or private nonprofit institutions of higher education, with at least 25 percent or more total undergraduate Hispanic full-time equivalent student enrollment.

> **BOX 3-1 Responding to Asian Growers in California**
>
> More than 700 farms in Fresno County, California, are operated by immigrant Asians, mainly Hmong, Mien, and Laotian farmers who arrived in the 1980s and 1990s as political refugees of the Vietnam War. Most of their farms are 10–15 acres, with intensive year-round farming and multiple specialty crops, including lemongrass, luffa, Chinese long beans, sugar peas, moqua, opo, bitter melon, and bok choy.
>
> The University of California Cooperative Extension has developed vegetable production techniques that help those and other small-scale farmers to raise crops more efficiently and profitably. These techniques include the use of plastic mulches to control weeds, drip irrigation, and plastic row covers. In 1998, extension personnel also piloted a biweekly call-in program on a local Hmong radio station on topics relevant for local farmers, many of whom are not fluent in English. Calls from Hmong farmers increased 300% during the first 6 months of the show and 800% in the last 6 months of the show (Ilic, 1992; Kan-Rice, 1999; USDA, 1999c).

only in response to considerable public pressure and to such litigation as the 1997 class-action lawsuit filed against USDA by African American farmers (Pigford v. Glickman, 1997). The committee encourages an even greater public-sector effort to engage these populations.

Recommendation 6
The public sector, at both the federal and the state level, should expand its programming focus with minority-serving institutions, which have unique access to underserved groups.

More effective communication with these groups would help research institutions move toward conducting research and extension that are relevant to their circumstances.

TECHNOLOGY TRANSFER

The previous section addressed heterogeneous producer characteristics that affect adoption of technology, and it proposed a public policy approach for responding to this heterogeneity. We now turn to a discussion of private- and public-sector technology transfer systems and the influence of public-sector technology transfer on the structure of agriculture. Technology transfer, defined in the broad sense, involves conveying information to the user and has traditionally been in the purview of Cooperative Extension and other areas within the public-sector system. Embedded within that definition, as a result of growth of private sector and patent protection legislation (U.S. Congress, 1980), is the protection of intellectual property. Strengthening of intellectual property rights over the past 20 years has had important effects on the technology transfer area and, consequently, important distributional consequences. This area is discussed further in Chapter 5.

Market-Oriented Technology Transfer

The private sector is increasingly important relative to the public sector in the delivery of agricultural technology. According to a survey conducted in 2001, more than 25 percent of agribusiness companies have total marketing budgets (including direct marketing, sales literature, farm shows, public relations, advertising, and internet marketing) exceeding $1 million dollars, and expenditures in agribusiness marketing budgets are increasing over 1998 levels (Agri-Marketing Association, 2001). In contrast, real aggregate federal expenditures for public extension have declined over time, from $332 million in 1991 to $280 million in 2000 (CSREES, Office of Extramural Programs).

Other surveys indicate that the private sector (e.g., input suppliers) is a significant source of technologic information. For example, in 1998, 38 percent of farmers using precision agriculture relied on input suppliers for advice compared to 17 percent of farmers who used the extension service as a source of information about the technology (Daberkow and McBride, 2001). In the 1998 Farm and Ranch Survey, 35 percent of farms relied on irrigation equipment dealers for information on water conservation and irrigation cost reduction compared to 41 percent who relied on extension agents or university specialists (USDA, 1999c). The 1991–1993 Area Studies survey data indicate that fertilizer company recommendations exceeded the extension service as a source of information on nitrogen fertilizer application decisions. The survey also shows that a greater percentage of pest management advice was provided by chemical dealers than by local extension for corn, cotton, potatoes, soybeans, and wheat (Caswell et al., 2001). Survey data from the 1993–1995 Chemical User Survey indicated that chemical dealers were also the most-used source of pest management information for a variety of fruit crops (USDA, 2000a).

The private sector contributes to the process of transforming an invention from discovery to application by investing additional resources for validation, manufacture, and distribution. In some cases, private sector groups provide additional technology support outreach in addition to or in concert with the public sector. New institutional and financing arrangements between public and private sectors for technology transfer are discussed later in this chapter. Private-sector vendors logically seek to work with farm operators who can contribute the most to their profits, and so they are more likely to seek out larger farm operations and develop products that make it easier to manage more land or animals with less labor.

Extension: Public-Sector Technology Transfer

The Cooperative Extension Service was established under the Smith-Lever Act in 1914 (U.S. Congress, 1914). Extension receives public support through CSREES and from state and local governments. A major function of the extension service has been to translate information from research for farmers and other citizens through adult education. State extension organizations provide administrative support and subject matter specialists. A local extension agent system with county offices throughout the nation was developed to produce and distribute information on applied problems. The Cooperative Extension Service is accountable to county governments, state governments, land grant universities, and the federal government through CSREES and by myriad grants awarded through other private, state, and federal entities.

STRUCTURAL IMPACTS OF EXTENSION

Cooperative Extension is an important link between research and the structure of agriculture. Huffman and Miranowski (1981) demonstrated that extension can be a substitute for formal education and human capital in adopting technology. Thus, extension can ensure that the structural implications of new technologies are considered. Extension can affect agricultural structure through *what* is communicated (or what is *not* communicated), *to whom,* and *how* that information is communicated.

Populations Targeted by Extension

For several decades there has been a substantial literature on the relationship between farm size (and other indicators of farmers' socioeconomic status) and contact with Cooperative Extension and other "change agents." The early literature is summarized in Everett M. Rogers' *Diffusion of Innovations* (1995). More recently, Huffman and Evenson (2001) showed that public R&D and education (including extension) have been at least as important as private R&D and market forces for changing livestock specialization, farm size, and farmers' off-farm work participation from 1953–1982 (discussed in Chapter 2). A survey relating herd size and five dimensions of contact with extension among Wisconsin dairy farmers reported that farmer use of extension services was highly variable and appeared to be correlated with the size of the farm (Ostrom et al., 2000).

TABLE 3-1 Contact with Extension by Herd Size

Type of Contact During Past Year*	Herd Size			
	1–49 Cows	40–99 Cows	100–199 Cows	>200 Cows
Read an extension publication or article	79	90	90	90
Visited county extension office	43	59	61	66
Called or spoke with extension agent	37	54	71	69
Attended extension meeting or workshop	22	41	50	48
Extension agent visited farm	14	22	37	62

*Percentage of operators within herd size category reporting any contact during the past year.

SOURCE: Adapted from M. Ostrom, D. Jackson-Smith, and S. Moon, 2000. Wisconsin dairy farmer views on university research and extension programs. Wisconsin Farm Research Summary. Summaries of Research from the Program on Agricultural Technology Studies 2 (January): 1–6.

As shown in Table 3-1, the disparity in extension contact by herd size is smallest for reading extension publications or articles but is very large in the case of the extension agent visiting the operator's farm[4]. The study also reported that, as herd sizes increased, farm operators were more likely to report that extension programs had been beneficial to their farm business. A disproportionately large share of the dairy farmers who were "unsure" whether they had benefited from extension came from farms with smaller herds. Using data from a 1989 survey of North Carolina farmers, Flora et al. (1993) found that several factors were associated with frequency of contact with extension agents and extension information: race (whites much more than African Americans or Native Americans), gender (men much more than women), size (large much more than small, measured either by acres or by gross sales), and type of agriculture (conventional more than alternative). Although some extension programs, such as Missouri's Small Farm Family Program, have worked with limited-resource and small farmers, extension's audience is composed mostly of larger-than-average farm operators.

---

[4]Note that for nearly two decades, large farmers have tended to rely more heavily on private information sources such as crop consultants and agribusiness field or sales representatives (for a recent overview see Wolf, 1998). Extremely large operations such as industrial-scale dairy farms and large cattle feedlots often hire their own technical staff (e.g., agronomists and veterinarians). However, the tendency for large farmers to pursue information sources other than the traditional county agent system does not generally reflect a decreased tendency for large farms to contact extension, much less a trend toward relegation of Cooperative Extension programs to smaller producers. In recent years, in fact, very large operators often contact land grant extension specialists directly as part of the diversification and intensification of their search for the latest information on production practices.

Some groups traditionally have received little support from extension. African American and Native American farmers have traditionally not fully accessed public-sector support in the area of technology transfer (USDA, 1997a; 1998b). When minority farmers are approached, the method may not be culturally or socioeconomically appropriate.

CHANGING THE FOCUS OF TECHNOLOGY TRANSFER PROGRAMS

The next section illustrates that the public-sector technology transfer system is changing through the development of partnerships with the private sector and through increasing engagement among public-sector institutions.

Partnerships with the Private Sector

Public-sector technology transfer is increasingly resulting from novel partnership arrangements with the private sector. Public-sector involvement has the potential to ensure that the results of research are publicly accessible. Mechanisms of technology communication are improving because of public sector networking with private (e.g., National Pork Board) and other public institutions. Technology transfer arrangements may include the following:

- *Universities invest in development and commercialization.* Some universities invest selectively in product development through partnerships with industry, specialists, and farm advisors. These team efforts can be partially financed through private-sector contracts or contributions. University teams have been involved in seed development and commercialization for wheat, cotton, and other agricultural crops. University task forces also have developed IPM and waste disposal strategies.
- *Individuals who made discoveries while working in the public sector engage in private development of a product.* University or USDA researchers sometimes start private businesses based on work conducted in the public sector. In other cases, public-sector employees serve as consultants to private enterprises. Many entrepreneurs commercialize the knowledge and findings they obtain while working toward graduate degrees. Some consulting companies established by university researchers and graduates offer managerial or agronomic expertise. In other cases, for example in the biologic control business, enterprises provide expertise and a product (beneficial insects).
- *Private-sector entrepreneurs and companies invest in development and commercialization of university findings.* The classical processes of

technology transfer are still common. Commercial entities develop technology innovations reported by government-supported researchers in the scientific literature, frequently enlisting public-sector researchers as consultants.
- *Private companies finance university research in exchange for the right to develop and commercialize the resulting innovation.* For example, Iowa State University's Pig Genome Mapping program is facilitated by private industry through provision of financial and genetic resources.
- *Private entities buy the rights to commercialize university patents or varieties.* In recent years, the use of formal processes of agricultural technology transfer has increased. For example, the University of California at Davis received more than a million dollars one year for the right to use its strawberry varieties. The chemical industry bought the patent rights for dibromochloropropane (DBCP) from the University of Hawaii to develop nematode control strategies.
- *Cooperative Extension forms partnerships with the private sector, particularly for the production of public goods.* For example, the Farm Bureau, Trees Forever, chemical companies, and Extension are collaborating to install riparian buffers in Iowa. Nonpublic sources of extension funding, including grants and contracts, often are used in technology transfer.

The structural implications of public-private sector partnerships are an important issue needing further analysis but are beyond the charge of this committee.

Partnerships Among Public-Sector Institutions

A recent report by the Kellogg Commission, *Returning to Our Roots: The Engaged Institution* (1999), endorses the concept of institutional engagement. The report's recommendations encourage institutions to go beyond conventional one-way outreach and service and to become "more sympathetically and productively involved with their communities" with a "commitment to sharing and reciprocity". As extension considers this model, its technology transfer activities should be more effective in reaching a more diverse audience, and that could have implications for the structure of agriculture.

Consistent with the model of engagement, the structure, function, and processes of extension are changing and may provide access to a wider set of expertise within the university community, engage farmers and others in the research process, and facilitate improved accessibility of options.

Changes in Extension Structure

Extension is increasingly positioned at a higher administrative level than the college of agriculture within many universities. In 1995, extension played a university-wide role outside colleges of agriculture in 44 percent of 186 land grant universities surveyed (Warner et al., 1996). A more recent survey found the same distribution for land grant universities (Luft, 2000). Examples of university-wide positioning can be found at the University of Wisconsin, Oregon State University, and Iowa State University. From this administrative vantage point, extension can access a wider range of university resources and expertise. An institution-wide arrangement has fostered new interdisciplinary research and more holistic outreach that broaden the range of users, expand the utility of research, and change the research agenda.

Changes in Extension Function

Extension is increasingly responding to problems that go beyond its traditional focus on agricultural production and farm programs. The private sector has focused on transferring technologies from which value can be captured in the production or sale of seeds, machinery, agrochemicals, and plant and animal nutrients. Extension consistently engages in higher risk endeavors, particularly those that involve alternatives or farm groups considered less profitable for the for-profit sector, such as the production of public goods, from which revenue cannot be captured.

Surveys in Missouri and focus groups in Minnesota have demonstrated the broad set of expectations for extension and needs identified by local stakeholders (Warner et al., 1996). CSREES has operated to connect other federal agencies with its state extension partners (Box 3-2).

A more broadly defined extension service in terms of constituencies and stakeholders can ensure that structural dimensions of research results and technologies are considered.

Recommendation 7
Extension should continue to reach out to other programs within universities, to draw wider networks of human resources, and to work with broader arrays of partners in the federal, private, nonprofit, and client sectors. CSREES should continue to facilitate more interdisciplinary and interagency activities that involve its state extension partners. CSREES should evaluate the potential and effectiveness of these extension approaches to serve diverse constituents.

> **BOX 3-2 State Extension Partners are Linking to Other Federal Agencies on a Broad Array of Problems**
>
> *Children, Youth, and Families at Risk Program and the U.S. Military*
>
> USDA's Children, Youth, and Families at Risk Program, established through a congressional appropriation in 1991, has forged links between Cooperative Extension and U.S. Military programs for children and families. The U.S. Army approached CSREES in 1995 for help in developing more comprehensive youth development programs on army installations. In 2001, 24 extension professionals on temporary assignment from their universities provided technical assistance in developing youth programs, including 4-H, for 130 installations. CSREES also has brokered a linkage between the U.S. Air Force and land grant universities to conduct research on preventing family violence. Cooperative Extension and U.S. Air Force bases in nine states are working together to identify critical needs of and build programs to serve at-risk children. In one state, this collaboration has developed a strong recreational component to help youth attain life skills. In another state, at-risk youth are using theater as a way to cope with issues of substance abuse, teen pregnancy, and violence.
>
> *Geospatial Technology Program and NASA*
>
> CSREES and the National Aeronautics and Space Administration signed a memorandum of understanding in 1998 to initiate three pilot studies in the application of geospatial and remote-sensing technologies to agriculture and natural-resource management. Permanent extension positions have been established and cofunded in three states.
>
> Partner institutions have conducted needs assessments to investigate the potential of different constituencies (e.g., ranchers and farmers) for adopting geospatial technologies. Tools have been developed to improve spatial literacy and technology access. Partner institutions have invested in progressive uses for the technology, such as economic development and conflict resolution. For example, disenfranchised, at-risk youth from minority communities in one state learned how to use Global Positioning System receivers and Arc View to collect data and construct maps describing community resources. The project has promoted dialogue among community leaders and at-risk youth.
>
> USDA and NASA have collaborated through a $7.5 million interagency program in the FY 2001 funding cycle of the Initiative for Future Agricultural and Food Systems program.
>
> *NEMO and NOAA*
>
> Nonpoint Education for Municipal Officials (NEMO) is an educational program for local officials that addresses the relationship of land use to natural-resources protection, with an emphasis on water quality. The program, developed at the University of Connecticut, has engaged three collaborative partners, Cooperative Extension, the Connecticut Department of Natural Resources Management, and Engineering, and the Connecticut Sea Grant Program, with support from a variety of state and federal agencies, including the Connecticut Department of Environmental Protection, CSREES, U.S. Environmental Protection Agency, NASA, the U.S. Fish and Wildlife Service, and the National Oceanic and Atmospheric Administration National Sea Grant College Program. NEMO has worked with almost two-thirds of the 169 municipalities in Connecticut, and a national NEMO Network of projects based on the Connecticut model has projects in 19 states. NEMO educational programs have catalyzed a variety of local actions to protect water resources, including changes to zoning and subdivision regulations, open-space planning and acquisition, and implementation of vegetative best management practices like grassed swales and pervious alternatives to pavement.

Changes in Extension Process

Stakeholder participation is growing in extension, and extension workers are changing to be more receptive to farmers' perspectives. The definition of "stakeholder" is broadening, too, although land grant universities are still somewhat hesitant to broaden the range of stakeholders because of the perceived increased transaction costs and the fear of alienating some client groups (Extension Committee on Organization and Policy, 1996).

Whereas the "demonstration plot" model of Extension generally engaged farmers whose larger operations were considered exemplary, a participatory model is likely to involve a broader variety of farmer circumstances along the research–extension continuum. In California, for example, extension workers are facilitating highly applied, on-farm research. The SARE program (see Box 3-3) demonstrates that extension processes are increasingly engaging stakeholder participation.

BOX 3-3 Stakeholder Participation and SARE

The regionally managed Sustainable Agriculture Research and Education (SARE) program demonstrates how stakeholder participation can be integrated into research and extension. Stakeholder participation is engaged at three levels: priority setting, project review, and project implementation.

Stakeholder participation in administration makes SARE unique among federal granting programs. A broad group, including producers, farm consultants, university researchers and administrators, state and federal government agency staff, and representatives from nonprofit organizations, serves on the regional administrative councils that provide overall leadership for the program; establish program priorities, goals, and objectives; and select projects for funding.

Stakeholders serve on the technical boards convened by each regional administrative council to review the technical quality and relevance of SARE proposals. For example, the 2000 North Central SARE technical committee included 10 reviewers from the private sector (mostly producers) and 10 reviewers from the public sector—researchers and extension personnel from universities, ARS, the Natural Resources Conservation Service, and the U.S. Environmental Protection Agency (USDA, 2000e).

At the project level, SARE program has integrated participatory elements in research and extension. Since 1992, SARE has offered a small-grants program for farmers and ranchers to run their own on-site research experiments.

SARE also offers, through its Professional Development Program (PDP), learning opportunities for agricultural extension and other field agency personnel. PDP activities in the Northeastern Region have helped extension and other agency personnel identify better ways to work with producers as colearners and facilitators (USDA, 1998d).

> **BOX 3-4 Fax-Based, Satellite Information Request System: Reaching Small and Part-Time Farmers**
>
> An initial grant from the USDA Small Farms Program tested the usefulness of a fax-based, satellite information-request system to address the changing circumstances of small and part-time farmers in North Carolina. The project was tested in three North Carolina counties where at least one agribusiness could be enlisted to house a fax information request center inside its business location. The fax machines were used to request information from Cooperative Extension. In one county, as many as 200 people visited each week to obtain "hot topic" information about plant disease control and other issues (Richardson et al., 1998).

Extension is increasingly supporting farmer-to-farmer networking, although that has long been the basis of technology transfer, for example, through field days and demonstration plots. Examples include extension's support of the group Practical Farmers of Iowa and work with farm stewardship groups in Minnesota and farmer marketing groups in Illinois. Alternative forms of outreach and engagement, including use of the Internet, also are resulting in greater stakeholder participation (See Box 3-4).

SUMMARY

The literature on adoption suggests that various producers and farm operations adopt innovations differently. Different degrees of adoption can be signaled by characteristics of producers or farm operations, such as farm size, regional differences in land quality, availability of human capital, producer age, and tenure arrangements. Some research innovations are more likely to be adopted by specific groups of producers, with structural implications. An approach to setting priorities for research, based on needs assessment of a variety of users, is proposed as an avenue for targeting heterogeneous producers and farm operations.

This chapter discussed the structural impacts of extension through the disproportionate support for specific farmer groups. The chapter also contrasted the structural dimensions of the conventional "technology transfer" model of extension with new models characteristic of more engaged institutions. These new models are characterized by increasing collaboration with the private sector, changes in extension's position within universities, a broadening of the extension mandate through linkages with other federal agencies, and greater stakeholder participation in setting priorities for research and extension activities. Research is needed to analyze the structural effects of these collaborative approaches.

4

Structural Impacts of Public Investment in Agricultural Research

This chapter analyzes the public research portfolio and its structural implications. It first reviews the array of public-sector responses to structural issues and provides examples that illustrate public-sector efforts to monitor and analyze structural trends, serve the needs of diverse constituencies, and understand the effects of drivers of structural change. Next, the chapter compares empirical data on allocation of research spending among various research categories between 1986 and 1997. This section draws broad conclusions about the distribution of investments in the portfolio and their changes over time, with particular attention to research investments likely to better serve diverse producers or to incur structural change (based on the analysis in Chapters 2 and 3). In-depth analysis of investments in environmental research provides an example of a public-sector research investment that is likely to serve producers outside mainstream agriculture. Finally, innovative funding mechanisms are described as possible avenues for addressing structural issues.

PUBLIC-SECTOR RESPONSES TO STRUCTURAL ISSUES

Structural and distributional issues have increasingly become focal areas for the public sector. The Federal Agriculture Improvement and Reform (FAIR) Act of 1996 (U.S. Congress, 1996) and the Agricultural Research,

Extension, and Education Reform Act (AREERA) of 1998 (U.S. Congress, 1998) highlight the importance of these issues. AREERA authorizes coordinated programs to improve the viability of small and medium-sized operations and to support minority-serving institutions. FAIR established a competitive-education-grants program for Hispanic-serving institutions, and it mandated representation by minority-serving institutions on the National Agricultural Research, Extension, Education, and Economics Advisory Board.

Public-sector research that responds to structural issues can be broadly categorized into three major areas (Box 4-1): research to monitor and analyze structural variables; research that serves needs of diverse constituencies; and research to further explain other drivers of structural change, including the influence of alternative policy instruments on structural change (Chapter 5). Box 4-1 lists general examples of public research efforts to address those issues.

BOX 4-1 Public-Sector Responses to Structural Issues

Research Monitoring Structural Change

- The U.S. Department of Agriculture (USDA) Economic Research Service (ERS) has developed a significant body of research on structural trends, including a new farm classification system that divides U.S. farms into mutually exclusive and more homogeneous groups. Much of this is included in a Farm Structure Briefing Room on the ERS web site, http://www.ers.usda.gov/briefing/Farm/Structure (Hoppe et al., 2000; Appendix D).

- As part of the agricultural census, the National Agricultural Statistics Service (NASS) collects data on farm size, farm number, operator characteristics, and farm ownership.

Responding to Diverse Needs

- In 1999, USDA awarded $9.6 million in grants for research, training, and education to implement Hazard Analysis and Critical Control Points (HACCP) and other food safety advancements. Of that, $1.35 million was targeted specifically to assist small meat-processing plants and small farmers (USDA, 1999d).

- USDA's Agricultural Marketing Service (AMS) has partnered with the Sustainable Agriculture Research and Education (SARE) program to provide producer-led alternative marketing research and demonstration grants.

- The AMS Federal-State Marketing Improvement Program (FSMIP) provides matching funds to state Departments of Agriculture and other state agencies to conduct marketing studies or assist in developing innovative approaches to the

Continues

Public-Sector…continued

marketing of agricultural products. Priorities for FY2001 included increasing the base of marketing research and services of particular importance to small-scale, limited-resource farmers and rural agribusinesses and direct-marketing opportunities for producers to respond to expanding consumer demands for products and value-adding services. A project awarded in FY2000 developed and assessed demand for locally manufactured fruit brandy and port products and examined the extent to which sales of fruit brandy and port could be expected to enhance the potential income of small Missouri-based fruit producers (USDA, 2001a).

- The AMS Direct Marketing Plan identifies USDA's role in supporting marketing opportunities for small farmers. Through this plan, AMS is conducting research on direct-marketing opportunities, including farmers' markets, pick-your-own farms, roadside stands, subscription farming, community-supported agriculture, and catalog sales (USDA, 1998c).

- ARS conducted a program-by-program evaluation for all of its 22 National Programs on small-farm-relevant research projects in response to a recommendation by the USDA National Commission on Small Farms. An *ad hoc* group composed of representatives from the USDA National Commission on Small Farms and the ARS National Program Staff developed the criteria for determining what part of ARS research is applicable to small farms. The report concluded that more than two-thirds of current ARS research has the potential to contribute to small-farm income-earning capacity and competitiveness (USDA, 2000b).

- The October 1999 issue of the Agricultural Research Service's (ARS), *Agricultural Research*, was dedicated to research projects relevant to small farmers and ranchers (USDA 1999b).

- Six ARS units in Georgia are collaborating with university scientists to begin an economic and environmental impact analysis project, Small Farm Survival Project for the Southern Coastal Plan (USDA, 2001e).

- ERS has conducted a comprehensive assessment of certified organic farming, marketing, and acreage by state and by commodity (USDA, 2001c).

- In August 2000, ERS presented research results, "Goals, Financial Success, and Small Farms", examining farmers' ranking of various goals from the 1995 Farm Costs and Returns Survey, at the American Agricultural Economics Association symposium, Successful Small Farms: How Do They Do It?

- A 1999 issue of University of California's magazine, *California Agriculture*, was devoted to small-farm issues (University of California, 1999).

Drivers of Structural Change

- The Program on Agricultural Technology Studies at the University of Wisconsin was established by the Wisconsin State Legislature in 1990 to conduct research and outreach on the effects of new technology and public policy on family farming. *Continues*

Public-Sector...continued
Research topics include structural change among Wisconsin farms and its effects on rural communities, and the structural effects of different public policies, such as NAFTA, property tax reform, land-use planning, and milk price supports.

- Several land grant universities cooperate to analyze the structural effects of technology, farm policy, and tax policy. For example, the Food and Agricultural Policy Research Institute at the University of Missouri, the Center for Agricultural and Rural Development at Iowa State University, and the Agricultural and Food Policy Center at Texas A&M University have a long-standing cooperative effort in which farm-level impacts are analyzed from the perspective of individual, representative farms. Regional models also are used to evaluate the impacts of technology changes.

- ERS studied the effect of the current federal tax code on farming. This is the first study to apply the ERS farm typology to tax data (Durst and Monke, 2001).

Recommendation 8
The public sector should continue to acknowledge the importance of structural change in agriculture. ERS and NASS should continue to monitor and analyze structural change and its causes.

AGRICULTURAL RESEARCH INVESTMENTS

This section documents the committee's analysis of the public-sector agricultural research portfolio. It includes a time-series comparison of agricultural research spending between 1986 and 1997 and an analysis of the 1999 investment portfolio subsequent to the 1998 reorganization of the Current Research Information System (CRIS). The committee relied heavily on CRIS, and the next section provides background on the choice of its use as a data set.

Current Research Information System

CRIS is the USDA's documentation and reporting system for continuing and recently completed research projects in agriculture, food and nutrition, and forestry. Information is reported to the CRIS database by USDA intramural research agencies, state agricultural experiment stations, state land grant colleges and universities, 1890s institutions, state schools of forestry, schools of veterinary medicine, and USDA grant recipients. A variety of funders, including federal, private, and state sources, support projects reported to CRIS. For example, in FY 1999, USDA funding supported about 37 percent of the total research reported to CRIS. State appropriations accounted for about 35 percent; nonfederal funding sources, including private-sector sources, accounted for about 16 percent; and agricultural research funded by other federal agencies accounted for about 12 percent (USDA, 1999f).

The committee acknowledges several limitations of the CRIS data set. First, there are institutions, including private-sector institutions, that do not report to CRIS that do perform agricultural research with public funds. Second, information reported to the CRIS is not always reliable, and the classification of research into categories can be misleading. Third, the database does not comprehensively report agricultural research funding from state or other federal sources, although it does include some agricultural projects supported by federal agencies other than USDA. Despite these limitations, CRIS is the only uniform, longitudinal database available in which data are disaggregated by funding source, institution performing the research, and research program area.

Public Research Spending, 1986 and 1997

The committee used CRIS data to compare the distribution of funds from various sources among research areas for two years, 1986 and 1997, chosen for data compatibility (USDA, 1986; 1997b). The data are categorized by research subject and research goal. The information is summarized in Tables 4-1 and 4-2. Table 4-1 compares the distribution of public funds for agricultural research by commodity in 1986 and in 1997, and Table 4-2 compares the distribution of public research funds by goal. The committee's analysis presents the summary of total public research funds (state and federal). Although the committee did not disaggregate the data at the state and federal levels here, general observations are offered about state-level funding as a percentage of total state resources in particular areas and state funding relative to federal spending within an area.

Although Tables 4-1 and 4-2 show data for just two years, they illustrate general trends from which we can deduce several patterns of allocation. Most funds were devoted to research on agricultural commodities—both plants and livestock. These subjects received 53.86 percent of research funding (sum for categories 3 through 7 in Table 4-1) in 1986, but that funding dropped to 50.61 percent in 1997. Commodity research is applied, and it tends to result in biologic and agronomic innovation (including new uses for agricultural products). Most of the expenditure, 37.33 percent in 1986 and 33.42 percent in 1997 (sum for categories 4, 5, and 6) went to research on field crops, dairy, beef, poultry, and swine. States also appropriated the largest share of total state resources to research on agricultural commodities. Relative to federal sources, state

TABLE 4-1 Historic Allocation of Public Research Funds by Commodity

Classification category	Percentage of total public funds[a]		Difference
	1986	1997	
1. Water, air, soil[b]	9.07%	10.91%	1.84%
2. Forests, wildlife, fish[c]	12.80%	15.07%	2.27%
3. Fruits, vegetables, ornamentals[d]	13.12%	13.22%	0.10%
4. Field crops[e]	19.59%	17.90%	-1.69%
5. Dairy and beef[f]	11.05%	9.26%	-1.79%
6. Poultry and swine[g]	6.69%	6.26%	-0.43%
7. Other animals[g]	3.41%	3.97%	0.56%
8. Equipment[i]	0.74%	0.38%	-0.36%
9. Economic, nutrition, marketing[j]	9.67%	9.80%	0.13%
10. Weed, seeds, plants[k]	6.48%	5.89%	-0.59%
11. Others[l]	7.38%	7.34%	0.04%
Total (percent)	100.00%	100.00%	0.00%
Total (nominal 1,000 dollars)	$1,764,129	$2,721,509	$957,380

[a]Total public funds represent the sum of USDA-appropriated funding, CSREES-administered funding, other USDA funding, other federal funding, and state appropriations. "Other nonfederal funding" (including self-generated funds, funding from industry grants or agreements, and miscellaneous funds) is not included in the total.
[b]Soil and land; water; watersheds and river basins; air and climate; recreation resources.
[c]Timber forest products; range; wildlife and fish.
[d]Citrus and tropical–subtropical fruit; deciduous and small fruits and edible tree nuts; potatoes; vegetables; ornamentals and turf.
[e]Corn; grain sorghum; rice; wheat; other small grains; pasture; forage crops; cotton; cottonseed; soybeans; peanuts; other oilseed crops; tobacco; sugar crops; miscellaneous and new crops.
[f]Beef cattle; dairy cattle.
[g]Poultry; swine.
[h]Sheep and wool; honeybees and other pollinating insects; other animals.
[i]Farm supplies and facilities; housing and equipment.
[j]Food; people as individuals; family members; farm as a business; socio-political organization; agricultural economy U.S.; agricultural economy foreign; farm cooperatives; other marketing, processing, and supply firms; marketing systems.
[k]Weeds; seed research; biologic cell systems; plants.
[l]Experimental design/statistical methods; invertebrates; microorganisms and viruses; animals (vertebrates); research on research; management; research equipment and technology; unclassified.

SOURCE: Adapted from FY 1986 unpublished tables and U.S. Department of Agriculture, 1997b. Selected CRIS Funding Summaries, FY 1997, Table C: National Summary USDA, SAES, and other institutions by commodity.
[Online] http://www.cris.csrees.usda.gov/star/cristin.htm.

TABLE 4-2 Historic Allocation of Public Research Funds by Goal

Research Problem Area	Percentage of total public funds[a]		Difference
	1986	1997	
1. Sustainable resources management[b]	14.23%	15.40%	1.16%
2. Disease control[c]	23.38%	24.60%	1.22%
3. Enhanced productivity[d]	32.86%	26.57%	-6.29%
4. Improved products[e]	9.20%	9.69%	0.49%
5. Improve marketing[f]	2.62%	2.57%	-0.05%
6. Expand export markets[g]	1.43%	1.01%	-0.42%
7. Improve health and nutrition[h]	7.11%	7.91%	0.80%
8. Assist rural Americans[i]	2.22%	2.09%	-0.12%
9. Community improvement[j]	6.95%	10.16%	3.21%
Total (percent)	100.00%	100.00%	0.00%
Total (nominal 1,000 dollars)	$1,764,129	$2,721,509	$957,380

[a]Total public funds represent the sum of USDA-appropriated funding, CSREES-administered funding, other USDA funding, other federal funding, and state appropriations. "Other nonfederal funding" (including self-generated funds, funding from industry grants or agreements, and miscellaneous funds) is not included in the total.
[b]Ensure a stable and productive agriculture for the future through wise management of natural resources.
[c]Protect forests, crops and livestock from insects, diseases and other hazards
[d]Produce an adequate supply of farm and forest products at decreasing real production costs.
[e]Expand the demand for farm and forest products by developing new and improved products and processes and enhancing product quality.
[f]Improve efficiency in the marketing system.
[g]Expand export markets and assist developing nations.
[h]Protect consumer health and improve nutrition and well-being of the American people.
[i]Assist rural Americans to improve their level of living.
[j]Promote community improvement including development of beauty, recreation, environment, economic opportunity, and public services.
SOURCE: Adapted from FY 1986 unpublished tables and U.S. Department of Agriculture, 1997b. Selected CRIS Funding Summaries, FY 1997, Table D: National Summary USDA, SAES, and other institutions by research problem area.
[Online] http://www.cris.csrees.usda.gov/star/cristin.htm.

sources for production agriculture accounted for half or more of the total funding in many categories of production agriculture (data not shown).

Funding for research on field crops decreased from 19.59 percent in 1986 to 17.9 percent in 1997. The field crop sector is expected to continue toward increased size of operations and declining number of producers.

Specialty commodities (fruit, vegetables, ornamentals, and specialty animals: categories 3 and 7) received 16.53 percent of total public funding in

1986 and 17.19 percent in 1997. State funding for research on specialty commodities accounted for about a fifth of total state resources, about half of the total funding allocated to specialty commodities from all public sources, and was concentrated in a small number of states (data not shown). Total spending on specialty commodities relative to total commodities (sum of categories 3–7) increased from 31 percent of commodity spending (16.53/53.86) in 1986 to 34 percent of commodity spending (17.19/50.61) in 1997. Markets for some of those commodities are less saturated, and demand is more elastic than for major commodities; thus, specialty commodities could provide expanding sources of earning and value added to farms.

Less funding went to research on poultry and swine than to research on dairy and beef cattle. The poultry and swine industries have become industrialized (much of their output is produced through contracting or by vertical integration), whereas dairy and especially beef producers have retained competitive structures. Some of the major integrators of poultry and swine (e.g., Purdue Chicken, Tyson) have their own research facilities; dairy cooperatives generally fund public-sector research.

Little public funding was allocated in either year for basic mechanical or chemical research.

Finally, a modest but growing share of the research budget (from 9.67 to 9.8 percent) was allocated to economics, nutrition, and marketing (category 9, Table 4-1). The research results produced in those areas generally could be useful for small and large producers alike; however, results tend to be used more by producers with larger farms and more education.

Table 4-2 compares the distribution of public research funds by research problem area in 1986 and in 1997. The data suggest that the major goal of agricultural research is to increase production. That goal includes disease control, enhanced productivity, and development of new products. We observed a reduction in the allocation of resources to improved production categories from 65.44 percent (sum of allocations for problem areas 2, 3, and 4) of the budget in 1986 to 60.86 percent of the budget in 1997.

Improved marketing of food and fiber both in the United States and abroad (categories 5 and 6, Table 4-2) received a modest share of the budget—4.05 percent in 1986 and 3.58 percent in 1997. Low prices and unfavorable market conditions are major problems in agriculture (Gardner, 1992), and more research in this area could improve the income and welfare of farmers. Although marketing research could benefit all farms, some will undoubtedly help small farms remain competitive. Nonetheless, larger farmers have a greater incentive to use the information.

Finally, modest shares of research funds went to rural development projects (categories 8 and 9, Table 4-2), which received 9.17 percent of funds

TABLE 4-3 Allocation of Public Agricultural Research Funds, 1999[a]

Topic	Area total	Subcategory total
Administration		0.24%
Soil	5.11%	
Water	2.30%	
Management of range and forest	3.85%	
Natural resources general	5.67%	
Natural resources and environment		16.93%
Plant production	17.02%	
Plant protection	15.12%	
Plants and their systems		32.14%
Animal production	12.76%	
Animal protection	10.57%	
Animals and their systems		23.33%
Engineering and support systems		2.22%
Food products	3.64%	
Nonfood products	2.52%	
Food and nonfood products: development, processing, quality, and delivery		6.16%
Economic markets and policy		6.98%
Human nutrition	4.32%	
Food safety	3.49%	
Human health	1.32%	
Human nutrition, food safety, and human health and well-being		9.13%
Family and community systems		1.44%
Total research support administration and communication		1.42%
National total (percent)		100.00%
National total (thousands of dollars)		$2,815,834.00

[a]Total public funds represent the sum of USDA-appropriated funding, CSREES-administered funding, other USDA funding, other federal funding, and state

appropriations. "Other nonfederal funding" (including self-generated funds, funding from industry grants or agreements, and miscellaneous funds) is not included in the total.
REVISED: Major topic areas shown in bold.
SOURCE: Adapted from U.S. Department of Agriculture, 1999f. Selected CRIS Funding Summaries, FY 1999, Table D: National Summary USDA, SAES, and other institutions by research problem area. [Online] http://www.cris.csrees.usda.gov/star/cristin.htm.

in 1986 and 12.25 percent in 1997. Relative to federal sources, state sources accounted for approximately half of the total in these categories. The increase in public funding could have structural implications, to the extent that alternative rural livelihoods could improve as a result of the research. Small, part-time farmers would benefit from the results.

Public Research Spending, 1999

A revised CRIS taxonomy was approved by the CRIS Enhancement Steering Committee in 1998 (USDA, 1999f). Table 4-3 shows the allocation of funds to research problems by major topic area and by subcategory.

The new structure allows useful insight about allocations both by crop and by objective. Production issues, including plants, animals, and economic markets and policy, received 62.45 percent of total funding. The balance went to areas that do not contribute directly to agricultural production but that address environmental, engineering, nutritional, and social concerns of the agricultural and food sector.

A small but substantial share, 16.93 percent, was allocated to projects on natural resources and the environment. Although engineering received only 2.22 percent of the total funding, more than a third of that went to environmental engineering topics, including waste disposal, recycling, and reuse.

A significant share, 14.35 percent, was allocated to development of new products, to studies of economic markets and policy, and to studies of family and community systems. The largest portions of economic funding were allocated to natural resources and environmental economics. Environmental, consumer, and community resource and development economics received about 2 percent of the total public funding. Production and business economics received only 0.8 percent of the total public funding. Nutrition received 9.13 percent of the funding, of which 3.49 percent was allocated to food safety; the rest went to studies of human nutrition and health.

In summary, production agriculture, a significant force in encouraging structural change, remains the dominant recipient of public research funding. Crop protection, supply increase, and the development of new products are related major targets. However, the share of research funding allocated to production agriculture decreased from 1986 to 1997. Of the research dedicated to commodity production, specialty commodities received an increasing share of the portfolio; field crops research drew less; and poultry and swine research decreased relative to research on dairy and beef cattle. Our analysis also indicates that the public sector spends only a small proportion—less than one percent—on development of mechanical innovations (0.74 percent in 1986 and 0.38 percent in 1997), which are more likely to benefit large farms than small ones. Research areas that are likely to benefit small and underserved farms and large farms alike—economics, nutrition and marketing, and specialty commodities—received increasing attention. Our analysis also demonstrates that research on resource conservation, rural development, and improved health and nutrition is increasing. Those areas are likely to be scale neutral, so they will benefit diverse constituencies equally. Changes are slowly occurring in the process of broadening the criteria for setting research priorities. The committee encourages the public sector to continue in this direction.

PUBLIC RESEARCH AND ENVIRONMENTALLY SUSTAINABLE ALTERNATIVE AGRICULTURE

One way to evaluate the structural implications of public-sector research is to consider its relative contributions to developing knowledge and basic components of technology, such as those that support environmentally sustainable alternative agriculture. These are of interest to small-scale farmers, organic farmers, and others outside the commercial mainstream. The extent to which publicly funded research supports environmental technology is an indicator of its support for these constituencies. Mainstream agriculture also is adopting many of those technologies in response to more stringent environmental regulations. Alternative technologies include biologic pest control and IPM strategies (as alternatives to the use of chemical pest controls); the use of symbiotic microorganisms, including nitrogen-fixing bacteria (as alternatives for chemical fertilizers); on-farm composting and biodegradation of organic wastes (as an alternative to dumping or disposal); soil conservation tillage (as alternatives to conventional tillage); and management-intensive rotational grazing (as an alternative to open grazing or confinement).

The second column of Table 4-4, which relies on the CRIS database[1], shows a large number of current USDA-funded intramural and extramural research projects on topics that are crucial for alternative agriculture.

The record of U.S. patents granted from 1975 through 1998 for most of

---

[1] In each area listed in Table 4-4, the numbers of current publicly funded research programs were searched by research area code number and by technology keyword in the online CRIS database.

TABLE 4-4 Selected Alternative Agricultural Technologies: Current USDA-Funded Projects and Total Patents Granted, 1975–1998, by Type of Organization

Alternative Agricultural Technology	Total number of research projects at USDA and Land Grant Universities in 1999 (CRIS Data[a])	Number and Share of Public-Sector Patents Granted (1975–1998): Universities and Public Research Institutions (Micropatent Data)	Number and Share of Private-Sector Patents Granted (1975–1988): Individuals, Private Firms, Corporations (Micropatent Data)
Biocontrol of plant pathogens	161	60 (37%)[b]	101 (63%)
Biocontrol of insects[c] Bt[d]	30	22 (11%)	186 (89%)
Biocontrol of insects[c] non-Bt	362	50 (30%)	116 (70%)
Biocontrol of weeds	146	38 (62%)	23 (38%)
Encapsulation and delivery technologies for biocontrol applications	35	36 (32%)	76 (68%)
Insect pest management	1044	51 (57%)	38 (43%)
Nitrogen fixation (including nitrogen-fixing bacteria)	390	13 (34%)	25 (66%)
Beneficial soil microorganisms and bioinoculants (not including nitrogen-fixing bacteria)	165	15 (27%)	40 (73%)
On-farm composting and biodegradation	604	N/A[e]	N/A
Conservation tillage and no tillage	492	N/A	N/A

TABLE 4-4
Continued

Alternative Agricultural Technology	Total number of research projects at USDA and Land Grant Universities in 1999 (CRIS Data[a])	Number and Share of Public-Sector Patents Granted (1975–1998): Universities and Public Research Institutions (Micropatent Data)	Number and Share of Private-Sector Patents Granted (1975–1988): Individuals, Private Firms, Corporations (Micropatent Data)
Intensive rotational grazing	121	N/A	N/A
Total Number of Research Projects of All Types, 1999	17,320	N/A	N/A
Overall Distribution of Public to Private Recipients of U.S. Patents in 1984[f]		(3%)	(97%)

[a] USDA FY1999 CRIS data (USDA, 1999f).
[b] Percentages of patents for each technology are in parentheses.
[c] Total percentages calculated from statistical reports by U.S. Patent and Trademark Office, 1975–1998.
[d] This category does not include any transgenic applications of *Bacillus thuringiensis*.
[e] Not applicable.
[f] 1984 was chosen as a representative, midrange year.

the alternative agricultural technologies (columns three and four of Table 4-4) illustrates the historic division of labor between the public and private sectors in developing technologies. The distribution of patents particularly demonstrates the large role of public-sector research in generating alternative agricultural technologies. Using the Micropatent database of front-page data for U.S. patents, technologies were searched by keyword and then expanded to include both cited and citing patents. Patent search results in each category were examined individually and inappropriate matches were discarded.

Table 4-4 suggests that USDA and the land grant universities are involved in the science of alternative agriculture, funding the science to create technologies that are more responsive to the needs of farmers outside the large-scale mainstream and making American agriculture more environmentally sound. The proportion of intellectual property created by the public sector in alternative technologies is generally an order of magnitude higher than a baseline proportion of all U.S. patents from 1984, the mid-sample year; only 3 percent of all U.S. patents in that year were assigned to the public sector. In some areas that are crucial to alternative agriculture, however, the percentage of technology assigned to the public sector is much higher. For example, the public sector has about 37 percent of *Bt* biocontrol patents, 62 percent of patents on biocontrol of weeds, and 34 of patents on nitrogen-fixation patents.

STRUCTURAL IMPLICATIONS OF RESEARCH FUNDING MECHANISMS

Research funding mechanisms are shifting the programmatic focus toward structural issues. The USDA National Research Initiative Competitive Grants Program (NRICGP) recently began to increase funding relevant to structural issues and to the research needs of small farms. In addition, new funding mechanisms, such as USDA's Fund for Rural America and the Initiative for Future Agriculture and Food Systems (IFAFS), now encourage multistate, multidisciplinary, and multifunctional (linking research and extension) activities. Working across state lines and across disciplines could be a way to mobilize research and extension processes that offer viable alternatives to more constituencies. It is important to recognize, however, that unless limited-resource farmers are specifically considered, cross-state or cross-disciplinary collaboration also can provide technology and processes of engagement that primarily benefit large farms. The next sections describe competitive-grants programs, and Box 4-2 provides examples of research for those programs that responds to structural issues.

Fund for Rural America

The Fund for Rural America, authorized under the 1996 FAIR act, was created to expand economic opportunities for rural Americans (U.S. Congress, 1996). Starting in 1997, one-third of the fund, $33.3 million annually over 3 years, was dedicated to research, education, and extension grants in the areas of international competitiveness, profitability, and efficiency; environmental stewardship; and rural community enhancement. Portions of the $33.3 million in discretionary funding were targeted to research, education, and extension programs. This included $4.5 million in technical assistance and training for an outreach program for socially disadvantaged Americans and $12.8 million in research, education, and extension programs for priority areas, including telecommunications research and research on counteracting concentration in the livestock sector. The 2001 program will award some $9.5 million to integrated research and extension projects that focus on preserving the economic viability in rural communities, tracking demographic changes and rural community innovations.

Initiative for Future Agriculture and Food Systems

IFAFS, authorized by Congress in 1998, is a competitive-grants program that gives priority to interdisciplinary, multistate, multi-institutional proposals integrating agricultural research, extension, and education. Distributional concerns related to the viability and competitiveness of small- and medium-sized farms are highlighted among the six priority programs. In FY 2000, 19 grants—representing 16 percent of the total funding of $120 million—were awarded in this category.

In the first round of IFAFS funding in 2000, 17 percent of the total funding went to projects (15 percent of all projects) that had the phrases "small farm" or "underserved population" in the title (USDA, 2000d). More than half of the IFAFS proposals are still oriented toward production and to commodity subject matter that does not specifically acknowledge structural issues (USDA, 2000d).

National Research Initiative Competitive Grants Program

In 1991, Congress created the National Research Initiative (NRI), an expanded competitive-grants program at USDA. The NRI was funded in FY 2000 at $119 million. In 1999, a goal was established to award up to $5 million of the NRI for small farm research projects (OFRF, 1999). A National Research Council panel also identified needed research on the effects of the changing farm and agribusiness structure that could be addressed by NRI (NRC, 2000a). Research relevant to small farms was funded by several NRI programs in FY 2000 and reported by USDA at $3.4 million. In its FY 2001 program description, NRI encourages research proposals that assess and evaluate "impacts of industrialization on industry structure and performance" and "impacts of public policy alternatives on industry structure" (USDA, 2000c).

BOX 4-2 Research Funding and Structural Change

Fund for Rural America

Many of the fund's competitive grants have been awarded for research on structural and distributional issues relating to small producers and disadvantaged groups (1997 funding cycle):

- Strategies were developed for market improvement, genetic improvement, value-added processing, and new product development for the Navajo wool and mohair industry.

Continues

- Training videos that promote food safety were developed for educating workers in small (10–500 employees) and very small (fewer than 10 employees) meat and poultry plants. Processing techniques, such as hot water and acid treatment, are affordable and effective for very small facilities, and they produce marked reductions in microbial contamination.

- The effects of integrator practices on contract poultry growers were examined from economic, sociologic, and legal perspectives. Using a survey of poultry growers, contract terms and practices were investigated. Federal and state law governing relationships between integrators and growers also was analyzed. Finally, educational materials were developed to explain the analysis to poultry growers involved in dispute resolution or arbitration.

Initiative for Future Agriculture and Food Systems

Many IFAFS competitive grants have addressed structural and distributional issues relating to small producers and disadvantaged groups (2000 funding cycle).

- An integrated research, extension, and education program investigated direct-marketing systems of small farms and worked with vendors and managers to explain, evaluate, and improve the performance of these markets as profitable outlets for small farms.

- Small, mid-size, and limited-resource farmers were targeted for adopting beef or forage enterprises as an alternative to tobacco cultivation. Participants enrolled in "cow college" programs to learn successful production techniques, rotational grazing management, artificial insemination techniques, farm records analysis, value-added marketing, and leadership development.

- A research project investigated which factors motivate commercial banks, the dominant guaranteed loan users, to lend to small, socially disadvantaged, or beginning farmers. Bank financial data and variables were analyzed to explain commercial bank decisions to guarantee loans and the volume they lend to farms.

National Research Initiative Competitive Grants Program

Research awards in the NRICGP 2000 funding cycle also addressed structural issues and small-farm needs:

- An experimental survey investigated the extent of and motivations for integration and coordination decisions as reflected through contracting behavior on the part of agricultural producers.

- A project used Geographic Information Systems technology, focus groups, and interviews to investigate the constraints and opportunities small farms face in enhancing viability in the context of the local food system.

Although these public-sector competitive grants programs account for only about 10 percent of all publicly funded agricultural research ($2.8

billion in FY1999), they increasingly address issues relating to small and medium-sized farmers, respond to concerns of underserved constituencies, and encourage a closer linkage of extension and research in multistate, multidisciplinary, and public and private efforts emphasizing research that bears on the structure of agriculture.

Recommendation 9
The public sector should continue to experiment with research approaches—including multifunctional partnerships that link research and extension, partnerships that link the public sector with the private and nonprofit sectors, multi-state cooperation, and multidisciplinary collaboration—as instruments for serving small farms, minority farmers, and other underserved producers. The public sector should evaluate the potential and effectiveness of these research approaches to serve these constituents.

SUMMARY

Public research is responding to a broadening of criteria for priority setting in research, which has implications for the structure of agriculture. This is occurring in three major areas: research to monitor and analyze structural variables; research to serve the needs of diverse constituencies; and research to further understand drivers of structural change other than research and development.

An analysis of the public-sector research portfolio demonstrates that although production agriculture still dominates, its share has decreased over time. Funding for research on specialty commodities, which offer opportunities for smaller growers to capture value, has increased relative to funding for research on other commodities, such as field crops. Support for chemical and mechanical research is minimal, whereas research on issues likely to benefit small and underserved farms as well as large farms—for example, natural resources and the environment, marketing, and rural development—is increasing. An analysis of research on environmentally sustainable technologies indicates that the public sector has played a major role in generating these technologies, many of them useful to farmers outside the commercial mainstream.

Innovative funding mechanisms integrating research and extension and fostering multidisciplinary research are suggested as possible avenues toward more effective investigation of structural questions.

5

Drivers of Structural Change, Changes in Knowledge and Information, Implications for Policy

Innovation results in change, and change almost invariably has a structural component. Because research and development (R&D) and technology transfer are important components and sources of innovation, it is not surprising that the activities and policies of the public and private sectors have structural impacts; the summaries of research in Chapters 3 and 4 clearly document many of them. This chapter first places R&D and technology transfer in context as drivers or determinants of structural change in the agricultural sector. The committee acknowledges that a discussion of drivers other than public-sector R&D is tangentially related to the charge of this study, but this context is critical to understanding the relative magnitude of impact of different drivers on structural change. Second, we briefly identify the characteristics of the agricultural sector of the future that are most likely to result from those fundamental drivers of change. Then, our discussion turns to important structural implications of the changing role of knowledge, information, and R&D that should be considered in the design of public-sector R&D and technology transfer policy. Finally, a set of research opportunities based on these arguments will be identified.

DRIVERS OF STRUCTURAL CHANGE

The U.S. food production and distribution industry is in the midst of major structural changes—changes in product characteristics, in worldwide production and consumption, in technology, in size of operation, and in geographic location. Productivity technology and public-sector R&D investments have been and will continue to be major determinants of comparative advantage and competitive position, including such considerations as public-sector support for research and technology transfer, the commercialization of new scientific discoveries, global trends and investments in new technology, and the status of intellectual property rights. However, R&D investments, technology, and innovation are only one component of many forces that drive change in agriculture. Other drivers contribute as well: pressures from consumers and end-use markets, changing demographics and work habits of U.S. families, changing attitudes about food safety and quality, increasing competition from global market participants, economies of size and scope in production and distribution, the inelastic characteristic of the demand for food[1], risk mitigation and management strategies of buyers and suppliers, strategic positioning, market power, and control strategies of individual businesses, and private sector R&D and technology transfer policies. Finally, the availability and cost of resources, including capital and finance, personnel and human resources, and information and industry infrastructure in general will significantly affect the future structure of the farming sector.

Relative Price of Labor and Capital

A critical interaction between resources and technology has occurred in the United States and in other places around the world. In the past, production agriculture in America was driven by technology to save physical labor. The cotton gin, steel plow, reaper, tractor, and combine harvester all conserved physical labor and increased efficiency. More recently, electronic and information technologies have been used to alleviate the scarcity of managerial labor and expand the size of business one manager can supervise.

The implications of technology and innovation for changes in the agricultural industry cannot be well understood without an appreciation for the concept of *induced innovation* (Hayami and Ruttan, 1985). According to the induced-innovation concept, a fundamental driver of R&D investments is the relative price of a resource—specifically capital or labor. R&D investments are

---

[1] The traditional literature on agricultural policy (Cochrane, 1993; Gardner, 1992; Schultz, 1964) argues that consolidation in agriculture results from a high rate of technologic change and low price and income elasticities of demand (where reduction in price or increase in income do not significantly alter quantity consumed). Technologic changes tend to increase supply and with low elasticities, the increase in supply results in significant reduction in farm prices. Thus technologic change and price elasticity have contributed to the falling relative prices of agricultural products and some of the resulting policy and structural issues.

focused on technology and innovation that will reduce the cost or increase the efficiency of the most expensive resources: Those resources that were more expensive before the innovation become more productive and less expensive. They are consequently used more in the production process. The essence of the argument is that resource prices and market forces encourage or induce R&D investments that result in changes in relative resource productivity, and when these changes result in substituting the less expensive resource for the more expensive resource, structural change occurs.

Hayami and Ruttan (1985) focused their analysis of induced innovation on the R&D investments in labor-saving, capital-using technology in the United States and contrasted that R&D investment with the labor-increasing, capital-saving investment in Japan. Their work documented that the high opportunity cost of labor relative to capital in the United States encouraged R&D and technology innovation in labor-saving machinery, equipment, and livestock facilities that can be most efficiently implemented by units of larger scale. In contrast, the relatively high price of capital and land relative to labor in Japan encouraged R&D investments that were labor and land intensive and better implemented by small-scale production units.

Studies by Kislev and Peterson (1981, 1982, 1996) bear out with the induced-innovation model, indicating that the high cost of labor in U.S. agriculture (in part because of attractive off-farm employment opportunities) has been seen in annual increases of 2–3 percent for labor costs and in the decreasing price of capital. The authors provide evidence that, during the late 1970s and early 1980s, when the price of labor relative to capital declined because of higher fuel costs, average farm size in the United States actually *declined* somewhat. Those data further support the hypothesis that the rising cost of labor relative to capital was the main cause of increase in farm scale. The logical economic response of a relative price change is to substitute capital for labor, and given the indivisibility of most capital items, it becomes more efficient to use capital-intensive technology on larger scale units. The consequence of this increase in the relative costs of capital and labor is more capital-intensive and larger scale farm units. An additional consequence is the incentive for R&D investments and technology transfer to further increase the efficiency of labor through innovations that increase its productivity—further reinforcing capital–labor substitution and growth in farm size.

The technology treadmill identified by Cochrane (1979) also has been important in American agriculture, and it has driven the adoption of much new technology and hence structural change. As new technology is introduced, the first few farmers to adopt the practice gain doubly. They increase the volume of their product and, in addition, gain revenue from market prices that largely depend on the old technology's production volume. There is a tremendous incentive to be an early adopter. As more and more farmers adopt a practice, supply increases, and prices fall. This forces the remaining farmers to adopt the new technology to increase production to compensate for lower prices. Thus, over time, the market drives farmers to adopt new technology if they wish to stay in farming. Since largest farmers are more likely to support the fixed costs of adoption of new technologies, they are most likely to adopt technologies early.

Knowledge and Information: A Changing Role

Farmers have long recognized the importance of education as a source of competitive advantage and continuous improvement in business and financial performance. Knowledge and information have always been important, but their relative importance has increased in recent years (Drucker, 1992; Peters, 1992). Whereas the physical resources of land, labor, and capital combined with some knowledge and information were the determinants of financial success in the past, the role of knowledge and information has and will likely become more important in the future for successful farm management. Superior knowledge and information will position farmers to use land, labor, and capital efficiently.

The system and mechanism by which farmers obtain new technology and information is changing dramatically. The number of private-sector providers of R&D and information is expanding relative to what is available from the public sector. Information is becoming more detailed with the potential for increased accuracy and resolution. Dissemination technology has reduced the cost of accessing information and will make real-time personalized messages available anytime and anywhere. Knowledge and information are becoming increasingly important drivers of control and structural change in agriculture, and access to information and intellectual property rights is an increasing source of conflict as information increases in value, and that value can be captured by the private sector.

Farmers now have access to more information from the private sector (for large-scale or integrated producers, or from internal sources) and less from the public sector. In many cases, providers of key farm inputs, such as veterinary pharmaceuticals and agricultural chemicals, also have become important suppliers of information, leaving the traditional extension service and land grant university–USDA (U.S. Department of Agriculture) complex at a significant disadvantage in providing the latest technology and information. Larger producers rate traditional public-sector information sources, such as county extension agents and university specialists, significantly lower than many other sources of information for production, marketing, or financial decisions (Ortmann et al., 1993). Privatization of information, R&D, and technology transfer often also restricts the access of scientists in the public sector to the latest scientific knowledge and advancements. The public-sector scientist's ability to test and verify the claims of private-sector providers or to further the scientific base of their own R&D activities is limited. These dramatic changes—

both in the importance of information and in who will be its preferred provider—raise questions about the changing role of the public sector.

Government Policy and Structure

Government policies other than R&D policy can profoundly influence the path of development in any industry. Policy that will shape the U.S. food system includes farm income support and risk mitigation (for example, crop insurance and disaster payment) programs; antitrust rules and the regulation of competition in the food system markets; international trade policy and agreements; public incentives and investments in technology transfer and the creation of knowledge; intellectual property rules and regulations; interest rate and tax policy; and regulation of food safety, the environment, worker safety, the transportation system, resource use, and conservation.

Studies of the direct influence of government policies on size and type of farms are limited and out of date. Analysis of tax policy in the 1980s indicated that, in the aggregate, tax burdens for farmers reflected the progressive rate structure of the tax code at that time but that taxes were generally lower and less progressive for farmers than they were for other taxpayers (Sisson, 1982). Specific agricultural industries, including the beef sector and specialty crops (tree crops and other perennials), provided significant tax-sheltering potential that was generally more advantageous to those with higher taxable incomes—whether that income came from agricultural production or from other sources (Carman, 1997; USDA, 1981). The tax deductions associated with capital investment (depreciation of equipment or debt interest) reduced the cost of capital relative to labor, encouraging capital-for-labor substitution and the use of larger production units when increased size enabled farmers to spread out the fixed costs of capital investment. A more recent Economic Research Service (ERS) analysis of the structural effects of Federal tax law reported that recent changes to Federal estate tax provisions will make it easier to pass farms to the next generation by exempting most small family farms from tax payment. The report also notes that the ability to transfer larger farms, combined with preferential treatment for farmland and other business assets, could, however, help to accelerate the trend toward fewer and larger farms (Durst and Monke, 2001).

There is disagreement about the structural implications of farm support programs. Using econometric analysis, Tweeten (1993) and Huffman and Evenson (2001) concluded that commodity programs did not affect farm number or farm size significantly in the long term. Other analysts argue that government payments favor larger farmers. A recent General Accounting Office (GAO) review of USDA's Agricultural Resource Management Study and Program Payments Reporting System found that in recent years, more than 80 percent of farm payments have been made to large- (gross sales of $250,000 or more) and medium-sized farms (gross sales between $50,000 and $250,000), while small farms (gross sales under $50,000) have received less than 20 percent of the payments. Because payments are generally based on volume of production, the average payment of small farms that received payments was much less. The portion of the payments that has gone to large farms has increased and the portion to small farms has decreased during the period from 1996 to 1999 (GAO, 2001). Similarly, an Environmental Working Group analysis of over 30 million subsidy payment records between 1996 and 1998 concluded that the flow of farm subsidies has favored large operations: 10 percent of the recipients collected 61 percent of the payments (Williams-Derry and Cook, 2000). Goetz and Debrtin (in press) argue that in counties where there are already exits from agriculture (which is part of farm consolidation), the amount of farm payments is correlated with the number of exits. In areas where exits have not begun—generally where there is more differentiation—farm payments are inversely correlated with exits from agriculture, suggesting that in these cases, farm payments decrease concentration. Data from ERS show that small farms (less than $250,000 in annual sales) receive 83 percent of the payments from conservation programs (Conservation Reserve Program, Wetlands Reserve Program, and the Environmental Quality Incentive Program; USDA, 1998a).

Regulatory policy has had variable distributional and structural effects by region. Analysis of the ban on the use of methyl parathion in some crops suggested significant effects among agricultural producers (Zilberman et al., 1991). Zilberman et al. (1991) studied the effect of banning methyl parathion use in lettuce and found that the overall effect on producers was not as significant as the effect on producers individually. Another study showed that the methyl parathion ban in apples and almonds would reduce consumer welfare because of higher prices but that the overall effect for producers would be relatively insignificant (Lichtenberg et al., 1988). There were, however, drastic differences in the effect of the ban within the producer sector. Although methyl parathion was an effective pesticide, many growers did not use it, and so they gained from the ban. Among pesticide users, in regions where pest control substitutes could be used, the impact of the ban was minimal. However, in two or three regions without pesticide alternatives, the elimination of methyl parathion resulted in a loss.

Public-sector R&D policy is not an exclusive driver of structural change in agriculture. Other factors, especially market forces and government policies other than R&D policy, are significant. To the extent that public and private R&D are heavily market driven, they have important structural implications. Private-sector R&D, which has grown in importance relative to public-sector R&D, focuses on value creation and on reducing the expense attributable to the highest cost resources. Larger farms will have more opportunities to capture value and reduce cost. Similarly, market forces that drive public-sector R&D also have structural implications.

CHANGES IN FARMING

Based on the drivers noted in the previous section, production agriculture will continue to face dramatic changes that have implications for the structure of agriculture and for the public agricultural research agenda. Agriculture is increasingly characterized by the changes discussed below (for additional detail, see Boehlje and Schrader, 1996; Boehlje, 1999; Tyner and Boehlje, 1997). All of them merit research.

Global Competition

Expanded market access is important to the future of global markets and international trade, but international transfer of capital and global access to technology and R&D are likely to be the most important dimension of more open trade. In the past, most private-sector technology transfer and R&D activity has focused on the United States and Western Europe. Today, these are relatively mature markets for R&D in terms of acreage growth and expansion of livestock production capacity. Growth opportunities for agricultural products are likely to be greater in Canada, Mexico, South America, Eastern Europe, and Asia. With the opportunities for global-oriented companies to expand their markets, one would expect substantial expansion in commercial technology transfer and R&D activity specifically focused on geographic regions outside the United States and Western Europe. The long-term consequences will be a narrowing of the gap between the productivity in those parts of the world versus traditionally dominant production regions and an increase in worldwide production capacity. This increased efficiency, productivity, and capacity in other production areas, along with the worldwide sourcing and selling strategies of global food companies, means that the United States and Europe might not be dominant players and that they will face increased competition in world markets.

Industrialized Agriculture

"Industrialized production" is large-scale production using standardized technology and management linked to the processor by formal or informal arrangements. Size and standardization are important characteristics in lowering production costs and in producing more uniform crop products and animals that fit processor specifications and meet consumers' food safety concerns and desires for specific product attributes. Smaller operations that are not associated with an industrialized system will have increasing difficulty gaining the economies of size and the access to technology required to be competitive, except perhaps in niche markets. Smaller operations can remain in production for longer periods, however, if they have less debt on facilities and are able to use family labor. Technologic advances, combined with continued pressures to control costs and improve quality, are expected to provide incentives for further industrialization of agriculture.

Differentiated Products

The transformation of crop and livestock production from commodity to differentiated product industries will be driven by consumer demand for highly differentiated food products, food safety, and trace-back ability for quality assurance; continued advances in technology; and the need to minimize total costs of production, processing, and distribution. Food systems will attempt to differentiate themselves and their products by science or through marketing. Scientific differentiation could include gaining exclusive rights to genetics through patentable biotechnologic discoveries, exclusive technology in processing systems, and superior food safety practices. Marketing differentiation could include branding, advertising, packaging, food safety, product quality, product attributes, bundling with other food products for holistic nutritional packages, and presentation of products in nontraditional ways. Based on analysis of the competitive success in ten leading trading nations, Porter (1998) makes a case for shifting from competitive advantage based on supplying lowest cost products to competitive advantage based on supplying differentiated products to sophisticated buyers.

Precision (Information-Intensive) Production

Production management is expected to move toward more micromanagement of specific production sites, spaces, and even acres or animals. The shift will be driven by the influx of information about environmental and biologic factors that affect production. The motivation for adopting micromanagement will be to minimize costs and enhance quality.

Increased use of monitoring technology, including sensors for individual monitoring and control systems, will greatly expand the amount of information available regarding what affects plant and animal growth and well-being. In addition, greater understanding of how various growth and environmental factors interact to affect biologic performance will be forthcoming. This understanding will then be integrated into management systems that incorporate the optimum combinations and apply them at a micro or localized level.

Ecologic Agriculture

In recent decades there has been an increased awareness of the importance of ecologic agriculture. Proponents argue that agriculture cannot function as an isolated system—one that has no exchange of matter or energy with its environment (Daley, 1996). They argue that agriculture must consider the limits of the natural resources used to produce commodities as well as the limits of the sinks needed to dispose of waste. In contrast to "therapeutic intervention" approaches, including chemical and biotechnologic pest management that lead to new problems because of the evolution of pest resistance, agroecologic approaches involve improving internal relationships in the system—improving predator–prey relationships, for example (Lewis et al., 1997; NRC, 2000b).

Some practitioners (notably biointensive integrated pest management operators and organic farmers) have made fundamental shifts in management practice by putting those principles into practice. In particular, they tend to use nutrient cycling instead of nutrient flows, self-regulating pest management systems instead of pesticide applications, and diverse crop–livestock systems instead of monoculture. Some practitioners have developed sophisticated production systems that have significantly reduced their energy input, substituted management skill for purchased input, and reduced their aggregate production costs. Whereas conventional approaches tend to be more capital intensive—they require the annual purchase of external inputs—agroecologic systems can require fewer capital outlays.

Ecologic approaches to land management are particularly relevant to the structure of agriculture, given that small farms (farms with less than $250,000 in annual gross sales) collectively hold 72 percent of U.S. farm assets, including 74 percent of land (measured in acres) owned by farms. Small farms thus can play a major role as stewards of natural resources and the environment, conserving collective public goods such as clean air, clean water, and biodiversity (USDA, 1999a).

Food Supply Chains

Managing and optimizing supply or value chains, from the genome to the consumer, will be increasingly emphasized. This supply chain approach will improve efficiency through better flow scheduling and use of resources; increase producers' ability to manage and control quality throughout the chain; reduce the food safety risk associated with contamination; and increase the ability of the crop and livestock industries to respond quickly to changes in consumer demand.

Food safety is a major driver in the formation of chains. One way to manage risk is to monitor the production and distribution process from genetics to final product. A trace-back system, combined with HACCP (Hazard Analysis and Critical Control Point) quality assurance procedures, can minimize the chance of contamination or quickly and easily identify sources of contamination. Trace-back may also be critical to implement identity-preservation systems and respond to consumers' concerns about food production processes and product characteristics.

A supply chain approach will increase interdependence among the various stages in the food chain; it will encourage strategic alliances, networks, and other linkages to improve logistics, product flow, and information flow. Future competition will not occur in the form of individual firms competing with each other for market share, but in the form of supply chains competing for their share of the consumers' food expenditures.

Increasing Risk

Agricultural production has always been risky, but it will be increasingly so in the future. Not only will the traditional variables, of price, weather, and disease, for example, continue to buffet the industry, new sources of risk are likely. Some food distribution channels could require particular quality characteristics that are not available in predictable quantities in open, spot markets. The risk of changing consumer preferences or a food safety scare could be much more difficult and important to manage than price or availability of raw materials. Unintended consequences of transgenic technologies may pose other new sources of risk. Contractual arrangements to obtain raw materials from a qualified supplier reduce price, availability, and contamination risks while ensuring predictable quality in the final product. However, this arrangement can reduce flexibility and introduce relationship risk—the risk that the qualified-supplier arrangement might be terminated.

The transformation of a segment of agriculture from a commodity industry to one that produces differentiated products introduces at least three new risks (Boehlje and Ray, 1999). First, differentiated products are positioned to respond to unique market segments that value the differentiated attribute. Assuming an attribute is measurable (which could be a risk in itself because many food attributes, including quality, are difficult to measure), consumers' and end-users' attitudes and willingness to pay for some attributes may change over time. For example, consumer attitudes with respect to food additives, biotechnology, and genetically modified organisms do not appear to be stable or predictable across cultures and time.

Second, alternative techniques to accomplish product differentiation could change, and the number of producers could increase. Thus, differentiated products are regularly commoditized over time, and initially high margins erode as new competitors appear. The rate of that process is also a source of uncertainty.

Finally, differentiated products in the food market, particularly branded products, also carry the risk as well as the reward of branding. Brand value can be destroyed quickly by defects or quality lapses. In food product markets, lack of food safety can destroy brand value quickly.

Increasing Diversity

Production agriculture in the future could be characterized by increasing diversity, which can overlap, but is different from, increasing diversification. Diversification involves expanding the number of activities or enterprises managed and controlled by one company. Diversity arises in the differences among the enterprises that constitute an industry. In fact, agriculture in the future could exhibit more specialization (less diversification) within a business but more diversity among businesses.

Agriculture in the past was characterized by typical or representative farms for various geographic regions, crops, or livestock products. Now, however, agriculture is characterized not by similarities among business entities, but by differences among them. Farms now produce corn and soybeans or hogs, whereas in the past one farm would produce all three products. Some farms specialize in breeding, gestation, and farrowing in pork production, and others specialize in finishing, the final feeding phase of pork production.

Diversity also increasingly characterizes the products of a segment of agriculture. With increasing diversity in consumer demands and with the opportunity for product differentiation at the production level, many farmers no longer produce commodity crop and livestock products exclusively. For example, some farmers produce high-oil corn, while their neighbors produce white or high-starch corn. Another source of diversity is the commitment to and dependence on farming as a source of family income. Many farm families combine farm employment with jobs in town or nonagricultural, home-based businesses.

Farming operations are now more diverse in size. Although large-scale businesses are growing rapidly in some parts of the livestock industries, smaller scale production units continue to be a significant part of agriculture. Smaller scale production frequently targets local customers (such as restaurants and higher income customers) and markets for specialty products, such as for premium hams produced without antibiotics or in free-range conditions. Diverse marketing and financial strategies also characterize those operations, including farmers' markets, roadside stands, farm-to-chef direct marketing, community-supported agriculture (CSA), and regional food systems, in which local producers and manufacturers provide food for a significant portion of a local population. In Iowa, for example, local producers and manufacturers have potential to provide food for a large portion of a local population. Practical Farmers of Iowa has helped to broker locally produced food for 47 different conference events at Iowa State University and at other locations in Iowa (Practical Farmers of Iowa, 2000). A hospital in Waterloo purchased $6,428 in local produce during the growing-season months—about 20 percent of the total produce it purchased. The hospital also purchased about $37,853 in locally produced meat in 2000. Before 1998, the hospital had purchased no local produce, and before 2000 it had purchased no local meat (Enshayan, 2000). CSA arrangements in Iowa have grown from 2 in 1995 to more than 50 in 2000 (Iowa State University, 2000).

Production technology adds an additional dimension of diversity. Some producers depend heavily on purchased inputs; others are more focused on sustainable production systems that recycle resources. Some farmers use highly capital-intensive production systems, whereas others who have more labor than capital find it more profitable to use labor-intensive technology and production systems. Thus there is increasing diversity in production technology, management and business practice, and financing and organization.

INFORMATION, INNOVATION, AND THE STRUCTURE OF AGRICULTURE

We have briefly reviewed the forces shaping the structure of the agricultural industry, including market forces, government policy, and innovation. We now turn more specifically to the structural implications of the changing role and sources of knowledge, information, and R&D. The discussion focuses on four dimensions or implications: structure and coordination, intellectual property rights and distributional consequences; globalization of information; and access to technology and the potential for disenfranchisement that should be considered in the design of future R&D policy.

Structure and Coordination

Many forces and drivers contribute to the structural changes in agriculture, but information and knowledge are particularly significant. As in other industries characterized by contractual arrangements, people who have unique and accurate information and knowledge have the power and control in the food production system that provide them capacity to profit from and transfer risk to the less powerful.

The increase in importance of knowledge and information for obtaining control, increasing profits, and reducing risk is occurring for two fundamental reasons: The food business has grown more sophisticated and complex so those with more knowledge and information about detailed processes and how to combine them into in a total system (a supply chain) will have a comparative advantage. The dramatic increase in information about the chemical, biologic, and physical processes of agricultural production will confer advantage to those who can put that knowledge to practical use.

In the past, production agriculture focused primarily on commodity products with coordination through open-access markets. The increased specificity in raw-material requirements, combined with the potential for producing specific attributes in agricultural products, is transforming part of the agricultural market from a commodity-product market to a differentiated product market. The need for greater diversity, more exacting quality control, and flow control will tax the ability of open markets to coordinate production and processing effectively. Open-access markets are a blunt instrument for conveying information about product attributes (quantity, quality, timing, etc.) and transaction characteristics (including services). Where open markets fail to achieve the needed coordination, other options—contracts, integration, joint ventures—will be used.

The speed of information flow and the rate of adoption with different coordination mechanisms are related to the difficulty in conveying information through open-access markets. In general, contract or ownership coordination results in more rapid transmission of information among the various economic stages. Consequently, the production and distribution system as a whole can react more quickly to changing consumer demands, economic conditions, or technology improvements. The ability to adjust rapidly is increasingly important because of the similarly rapid changes in economic and social systems worldwide.

The ability to respond quickly to changes in the economic climate is critical to maintaining profit margins. Likewise, it is essential to recognize poor decisions quickly and to make appropriate adjustments. A market-coordinated system characterized by biologic lags (e.g., a poor harvest) cannot respond to changing conditions as quickly as can an integrated or contract-coordinated system. That is, the response at one stage of a market-coordinated system can be initiated only after a price change signals a need. With little flexibility for adjustment during the growing and maturing processes, the change in quantity or quality is observed only after a full production cycle. By their nature, contract- or ownership-coordinated systems require more frequent and direct communication among the decision makers at each stage on a wider variety of product and service characteristics than is typically possible with more traditional open-access markets. The improved flow of information and more rapid adoption and adjustment allow contract- or ownership-coordinated systems to function more effectively in rapidly changing markets.

The logical question for individuals in the food manufacturing chain is how to obtain access to knowledge and information. Particularly for independent producers, knowledge and information are obtained from public sources and from commercial sources—genetics companies, feed companies, building and equipment manufacturers, packers, and processors. In general, independent producers obtain knowledge and information much the same way as they obtain physical and financial resources and inputs. In contrast, in contract- or ownership-coordinated systems of manufacture, in which production, processing, and distribution are completely integrated, knowledge and information come from a combination of internal and external sources. Many of these enterprises or alliances of enterprises have internal R&D staffs who enhance the knowledge and information base. The information they obtain frequently is proprietary, and so it is not shared outside the enterprise. Control over proprietary knowledge confers strategic competitive advantage.

R&D in contract- or ownership-coordinated systems is more focused on total system efficiency and effectiveness than it is on individual components of the system. It is more efficient to integrate the nutrition, genetics, building and equipment design, health care, and marketing strategy than it is to address those areas separately. In addition to more effective R&D, such alliances or integrated businesses can implement new technology more rapidly over a larger volume of output to obtain a larger volume of innovator's profits. In the event that a new technology proves defective or an experiment fails, contract- or ownership-coordinated systems generally have monitoring and control procedures to detect deteriorating performance earlier and make adjustments more quickly than will market-coordinated systems.

As knowledge and information become more important sources of competitive advantage, those who have access will be more successful than those who do not. Given the declining public-sector funding for R&D and information dissemination, the expanded capacity of integrated systems to generate and adapt proprietary technology enables the participants in that system to more regularly capture innovator's profits at the same time that they increase control and reduce risk. This provides a formidable advantage to the contract- or ownership-coordinated production system over the system of independent stages and decision making.

Intellectual Property Rights and Distributional Consequences

Patent legislation, court decisions, and U.S. Patent and Trademark Office (PTO) rulings since the 1980s have dramatically changed the setting in which intellectual property rights must be considered (NRC, 1997a). Plant and animal innovations were unprotected by the original Patent Act in 1790. Since then, numerous changes have occurred to extend the range of intellectual property protections. Of particular importance was the 5–4 1980 U.S. Supreme Court decision in *Diamond v. Chakrabarty*, which opened the door to patenting genetically engineered organisms under the original Patent Act. Since that ruling, PTO has further interpreted the Patent Act to include new plants, seeds, germplasm, and nonhuman animals as inventions. Numerous acts of Congress and presidential orders, notably the Government Patent Policy (Bayh-Dole) Act of 1980 (U.S. Congress, 1980) and the Federal Technology Transfer Act of 1986, (U.S. Congress, 1986), have mandated patenting of federally funded endeavors and the promotion of technology transfer.

Until the 1980s, most publicly funded research that resulted in information and knowledge provided to producers was in the public domain. Since the 1980s, however, suppliers, consultants, and service firms increasingly gather data for production agriculture; the private sector plays a larger role in providing data, knowledge, and information; and private property rights have replaced common property concepts. Private property rights enable individuals who have those rights to capture value—to extract profits or payment from those who use property. Consequently, with the growing privatization of the knowledge and information markets, intense debates and litigation have occurred over the intellectual property rights to these resources, property rights to data, and control of data accessibility. Differential values based on the exclusivity or other dimensions of property rights can affect who will receive the most valuable information. For example, in smaller farm operations the value of data might be much lower than is the cost of collecting the data, so there is no incentive to collect or analyze information. Smaller farms therefore are at a disadvantage as privatization of knowledge and information increasingly favors larger businesses that can capture relatively more profit from the property rights in knowledge, data, and information.

If the public role in providing data and information continues to decline and private-sector activity continues to increase, there will be three distributional consequences. First, a major purpose of public information and data services historically has been to provide open access to potential users, irrespective of size or other characteristics. Expanded private-sector activity in the information markets would result in more of the information being provided at a profit instead of for the common good. Thus, knowledge and information access will generally become less open as public sources decline in relative importance (or begin to exhibit profit-seeking behavior the way private-sector providers do).

Second, because profit-seeking behavior is an important determinant for private-sector knowledge and information providers, those of the target audience who can and will pay the most will obtain the most and best information. One would expect that the largest, most sophisticated, and most specialized companies could pay more and would receive more attention by private-sector knowledge and information providers. The less affluent enterprises would receive less information—and profit less from it. Finally, private-sector information providers would extract payment or capture profits, thus redistributing revenues from the production sector to the service sector. Note, however, that if the information increases efficiency and adds value, it could bolster the incomes of the information provider and the producer alike—depending on the cost to the buyer extracted and how incremental revenue is shared.

With growing privatization of knowledge and information, public-sector providers will increasingly face scrutiny about access to their information, the constraints they place on availability, and the audiences they target (who gets the information and at what cost). Growing concern about benefits and the economic and political power conferred by differential access to information will fuel this questioning.

Global R&D and Information

At the same time that knowledge and information are becoming more critical resources for success in production agriculture, globalization is fundamentally changing the nature of competition in the agricultural industry. During the 1970s and 1980s, two critical changes occurred: Public-sector and private-sector investments increased in almost all geographic regions of the world, and more technology and innovations were shared across national borders through public-sector international research centers and internationalization of agribusiness (Pardey, 1992). Globalization of agricultural research and development in technology contributes significantly to increased international competitiveness in agricultural product markets; no longer does one country or region of the world have exclusive and unique access to the latest information or technology with respect to genetics, nutrition, veterinary management, or pest control, for example.

The combination of globally adaptable production technology with site-specific information on soils and climatic conditions has added to the intensity of international competition. Information, as noted earlier, is increasingly a source of competitive advantage, and it is now being acquired and transmitted globally. The significant and profound implications are that internationalization of information and technology markets contributes further to international sourcing of products by agribusiness, international distribution of inputs by suppliers, and generally increased global competitiveness in the agricultural sector (Pray, 1993). A logical and yet largely unresolved public policy challenge involves distribution of international intellectual property rights.

Access to Technology and Disenfranchisement

The privatization of agriculture R&D and information markets, the profound structural changes occurring in the food production and distribution industry, and the narrowly defined criteria for allocating public-sector R&D funding all have the potential to restrict the access of some producers to the latest technology and innovation. Privatization also can block access to technology and R&D—even to those in the public sector. For example, if private-cost-driven, productivity–efficiency criteria are used for the selection and assessment of public-sector R&D activities, activities that might emphasize value-added production for producers would not fare well. Neither will sustainable-production practices that consider public as well as private cost. Nor will R&D focused on maintaining diversity to reduce risk if that diversity requires giving up some efficiency and incurring cost. Research focused on unique technologies of small-scale producers and labor-intensive operators also is not likely to be funded with the narrowly defined productivity–efficiency criteria of evaluation or assessment. The privatization of knowledge, information, and R&D; the induced structural change that results from that privatization; and the narrowly defined criteria for assessing and evaluating public-sector R&D have significant implications for producers. For an increasingly large number of producers, those factors will result in disenfranchisement and in restricted access to public- and private-sector innovation and R&D.

The implications of increasing diversity in the farm sector also are important in terms of access to innovation and new technology. As each farm operation becomes increasingly different from its neighbor, the knowledge, information, and R&D needs of all farms will diverge. Increased diversity requires information and technology providers to design products and services for individual customers or producers. This goes beyond the well-recognized rule that "one size does not fit all"; now, "one size fits only one or at the most very few" customers or information users.

The complexity of serving an increasingly diverse industry points to the need for more sophisticated service and information delivery systems to replace ineffective mass distribution systems. The complexity also increases the likelihood that some segments of the farm population will be underserved or excluded from the knowledge, information, or inputs they need to compete efficiently and effectively. Increased diversity could contribute to the disenfranchisement of some segments of the farm population and cause conflict among others. The farming population today shares little commonality of interest, objectives, and understanding based on common experiences, and it competes more than ever before for limited resources to meet the demands of different customers, constituencies, and clienteles. Increased diversity poses a significant challenge to those who want to provide knowledge, information, and technology to the production sector, and to those who want to represent the production sector in the shaping of public policy, including farm programs and public-sector R&D policy. In essence, the increased diversity in production agriculture results in increasingly diverse demands with respect to public-sector assistance or support for the industry.

RESEARCH OPPORTUNITIES

Based on the arguments articulated in the previous sections, the committee developed the following list of research opportunities relating to drivers of structural change:

- Research is needed to better explain the market forces that drive structural change and the specific influence of these market forces on consolidation (the number and size of farms, processors, input suppliers, and retailers), and on vertical coordination between various stages of the food production and distribution value chain.
- Research is needed on government policies that drive structural change and their specific consequences for consolidation (the number and size of farms, processors, input suppliers, and retailers) and for vertical coordination among various stages of the food production and distribution value chain.
- Research is needed on the implications of the transformation of agriculture from a market-coordinated commodity industry to a more tightly aligned, vertically coordinated, differentiated-product industry for the consolidation of production and distribution enterprises in the industry, the size and structure of those companies, the distribution of risk and returns they experience, and the potential for market power to result in monopolistic control or profits in the food industry.
- Research is needed to explain the implications of the privatization of knowledge and the expanding use of intellectual property rights for incentives to innovate, the distribution of costs and benefits from innovation, access to R&D, and consolidation and coordination of the agricultural production and distribution system.
- Research is needed to assess the implications of increasing global access to the latest information and technology from public- and private-sector R&D and technology transfer activities for the competitive and comparative advantage of U.S. farmers and the food production–distribution system, as well as for global consolidation and vertical coordination of input supply, production, processing, and food-retailing businesses.

SUMMARY

Many forces other than public-sector R&D policy affect the structure of agriculture. Those forces include the relative prices of labor and capital, the changing role of knowledge and information, and public policy. The structural implications of public-sector R&D and innovation policy should not be ignored, although it is likely that significant structural changes will occur in the agricultural sector irrespective of the structural bias or neutrality of public R&D

policy. The privatization of the R&D and innovation processes, combined with the increased diversity in the industry, raise legitimate concerns about access to the latest and best technology for all industry participants, regardless of size, business model (independent versus contract), or other structural characteristics. Consequently, a public-policy response to increase access to technology, target disenfranchised groups, serve a broader constituency, and evaluate (as well as include as part of funding criteria) the structural impacts of R&D investments is appropriate.

References

Agri-Marketing Association. 2001. Annual Marketing Survey. St. Louis, MO.

Alston, J. M., G. W. Norton, and P. G. Pardey. 1995. Science Under Scarcity: Principles and Practice for Agricultural Research Evaluation and Priority Setting. Ithaca, N.Y.: Cornell University Press.

Ashby, J. A., and L. Sperling. 1995. Institutionalizing participatory, client-driven research and technology development in agriculture. Development and Change 26:753–750.

Batte, M. T., E. Jones, and G. D. Schnitkey. 1990. Computer use by Ohio commercial farmers. American Journal of Agricultural Economics 72(4): 935–945.

Berardi, G. M. 1984. Socioeconomic consequences of agricultural mechanization in the United States: Needed Redirections for mechanization research. Pp. 9–22 in G. M. Berardi and C. C. Geisler, eds. The Social Consequences and Challenges of New Agricultural Technologies. Boulder, Colo.: Westview Press.

Bessant, K. C. 2000. Part-time farming situations among Manitoba farm operators: A typological approach. Canadian Journal of Agricultural Economics 48(3): 259–77.

Boehlje, M. 1992. Alternative models of structural change in agriculture and related industries. Agribusiness 8(3): 219–23.

Boehlje, M. 1999. Structural changes in the agricultural industries: How do we measure, analyze, and understand them? American Journal of Agricultural Economics 81(5): 1028–1041.

Boehlje, M., and J. Ray. 1999. Contract vs. independent pork production: Does financing matter? Agricultural Finance Review 59:31 42.

Boehlje, M., and L. F. Schrader. 1996. Agriculture in the 21st Century. Journal of Production Agriculture 9(3): 335–340.

Breimyer, H. F. 1991. Is there a family farm anymore? Pp. 3–9 in Rural Wisconsin's Economy and Society: The Influence of Policy and Technology. Madison, Wis.: Agricultural Technology and Family Farm Institute, April 10.

Brokensha, D., M. Warren, and O. Werner, eds. 1980. Indigenous Knowledge Systems and Development. Washington, D.C.: University Press of America.

Brumm, M. C., J. D. Harmon, M. S. Honeyman, J. Kleibenstein, and J. Zulovich. 1999. Hoop Structures for Gestating Swine. AED-44. MidWest Plan Service.

Busch, L., and W. B. Lacy, eds. 1986. The Agricultural Scientific Enterprise: A System in Transition. Boulder, Colo.: Westview.

Busch, L., J. L. Silver, W. B. Lacy, C. S. Perry, M. Lancelle, and S. Deo. 1984. The Relationship of Public Agricultural R&D to Selected Changes in the Farm Sector. Report to the National Science Foundation. Lexington, Ky.: Department of Sociology, Agricultural Experiment Station, College of Agriculture, University of Kentucky.

Busch, L., W. B. Lacy, J. Burkhardt, D. Hemken, J. Moraga-Rojel, T. Koponen, and J. de Souza Silva. 1995. Making Nature, Shaping Culture. Lincoln, Neb.: University of Nebraska Press.

Buttel, F. H., D. B. Jackson-Smith, and S. Moon. 2000. A Profile of Wisconsin's Dairy Industry, 1999. PATS Research Summary No. 3. Madison, Wis.: Program on Agricultural Technology Studies, University of Wisconsin.

Carpenter, J., and L. Gianessi. 1999. Herbicide tolerant soybeans: Why growers are adopting Roundup ready varieties. AgBio Forum 2(2): 65–72. [Online] http://www.agbioforum.missouri.edu/archives.htm.

Carman, H. F. 1997. U.S. Agriculture Response to Income Taxation. Ames, Iowa: Iowa State University Press.

Carr, P., G. Carlson, J. Jacobson, G. Nielson, and E. Skogley. 1991. Farming soils, not fields: A strategy for increasing fertilizer profitability. Journal of Production Agriculture 4(1): 57–61.

Caswell, M. 1991. Irrigation technology adoption decisions: empirical evidence. Pp. 295–312 in The Economics and Management of Water and Drainage in Agriculture, A. Dinar and D. Zilberman, eds. Norwell, Mass.: Kluwer Academic.

Caswell, M. F., and D. Zilberman. 1986. The effects of well depth and land quality on the choice of irrigation technology. American Journal of Agricultural Economics 68(4): 798–811.

Caswell, M. F., D. Zilberman, and G. E. Goldman. 1984. Economic implications of drip irrigation. California Agriculture. 38(7–8): 4–5.

Caswell, M., K. Fuglie, C. Ingram, S. Jans, and C. Kascak. 2001. Adoption of agricultural production practices: Lessons learned from the U.S. Department of Agriculture Area Studies Project. Economic Research Service Agricultural Economic Report No. 792. U.S. Department of Agriculture, Washington, D.C. [Online] http://www.ers.usda.gov/publications/aer792/.

CGIAR (Consultative Group on International Agricultural Research). 1999. Crossing perspectives: Farmers and scientists in participatory plant breeding: Cáli, Colombia: Program on Participatory Research and Gender Analysis, the Consultative Group on International Agricultural Research, 1999.

Chambers, R. 1983. Rural Development: Putting the Last First. London: Longman.

Cochrane, W. W. 1979. The Development of American Agriculture: A Historical Analysis. Minneapolis, Minn.: University of Minnesota Press.

Cochrane, W. W., 1993. The Development of American Agriculture: A Historical Analysis. Second Edition. Minneapolis, Minn.: University of Minnesota, Minneapolis.

REFERENCES

Cohen, S., A. Chang, H. Boyer, and R. Helling. 1973. Construction of biologically functional bacterial plasmids in vitro. Proceedings of the National Academy of Sciences USA 70: 340–344.

Daberkow, S. G., and W. D. McBride. 2001. Information and the adoption of precision farming technologies. Selected paper for presentation at the 2001 American Agricultural Economics Association meetings, August 5–8. Economic Research Service, U.S. Department of Agriculture, Washington, D.C.

Daley, H. 1996. Beyond Growth. Boston: Beacon Press.

Danbom, D. B. 1986. Publicly sponsored agricultural research in the United States from an historical perspective. Pp. 107–131 in New Directions for Agriculture and Agricultural Research: Neglected Dimensions and Emerging Alternatives. K. A. Dahlberg, ed. Totowa, NJ.: Rowman & Allanheld.

DeJanvry, A., P. LeVeen, and D. Runsten. 1980. Mechanization in California Agriculture: The Case of Canning Tomatoes. Berkeley: University of California Press.

Diamond v. Chakrabarty. 1980. 447 U.S. 303.

Drabenstott, M. 1998. This little piggy went to market: Will the new pork industry call the heartland home? Pp. 79–97 in Economic Review, Third Quarter. Kansas City, Mo.: Federal Reserve Bank of Kansas City.

Drucker, P. 1992. For the Future: The 1990s and Beyond. New York: Dulton.

Durst, R., and J. Monke. 2001. Effects of Federal Tax Policy on Agriculture. Agricultural Economic Report No. 800. Washington, D.C.: Food and Rural Economics Division. Economic Research Service, U.S. Department of Agriculture.

Dyer, J. 1999. In the Eye of the Stakeholder. Inquiry in Action. Madison, Wisc.: Consortium for Sustainable Agriculture Research and Education, B. Miller and K. Leval, eds. Newsletter No. 23.

Enshayan, K. 2000. Personal communication. University of Northern Iowa.

Extension Committee on Organization and Policy. 1996. The State Cooperative Extension Service and State Agricultural Experiment Station System Analysis of the National Research Council Report on "Colleges of Agriculture at the Land Grant Universities".

Feder, G., and G. T. O'Mara. 1981. Farm size and the adoption of green revolution technology. Economic Development and Cultural Change 30(October): 59–76.

Feder, G., R. E. Just, and D. Zilberman. 1985. Adoption of agricultural innovations in developing countries: A survey. Economic Development and Cultural Change 33(2): 255–298.

Federal Register. 2001a. Request for Proposals: Fund for Rural America, FY 2001. 20 April. 66(77): 20352–20365. [Online] http://frwebgate.access.gpo.gov/cgi-bin/getdoc.cgi?dbname=2001_register&docid=01-9745-filed.pdf.

Federal Register. 2001b. Request for Proposals (RFP): Initiative for Future Agriculture and Food Systems, FY 2001. 66(37): 11487–11509. [Online] http://www.reeusda.gov/1700/programs/IFAFS/rfp-2001.pdf.

Feldstein, H. S., and S. V. Poats. 1989. Working Together: Gender Analysis in Agriculture. Vols. 1 and 2. West Hartford, Conn.: Kumarian Press.

Fernandez-Cornejo, J., C. Greene, R. Penn, and D. Newton. 1998. Organic vegetable production in the U.S.: Certified growers and their practices. American Journal of Alternative Agriculture 13(2): 69–78.

Flora, J., C. Flora, and M. Schulman. 1993. The Impact of Family Labor Variables and Gender and Race of Farms Operator on Well Being of North Carolina Farmers. Chattanooga, Tenn.: Southern Sociological Society.

Forastieri, V. 1999. The International Labor Organization Programme on Occupational Safety and Health in Agriculture. [Online] http://www.ilo.org/public/english/protection/safework/agriculture/agrivf01.htm.

Friedland, W. H., and A. E. Barton. 1975. Destalking the Wily Tomato. Research Monograph No. 15. Davis, Cali.: Department of Applied Behavioral Sciences, University of California.

Friedland, W. H., A. E. Barton, and R. J. Thomas. 1981. Manufacturing Green Gold. New York: Cambridge University Press.

Fuglie, K. N., Ballenger, K. Day, C. Klotz, M. Ollinger, J. Reilly, U. Vasavada, and J. Yee. Agricultural research and development: Public and private investments under alternative markets and institutions. 1996. Agricultural Economics Report No. 735. May. U.S. Department of Agriculture, Economic Research Service.

Fulton, M., and Keyowski, L. 1999. The producer benefits of herbicide resistant canola. AgBioForum 2(2): 85–93.

Gardner, B. L. 1992. Changing economic perspectives on the farm problem. Journal of Economic Literature 30: 62–101. Reprinted in G. H. Peters, ed. 1995. Agricultural Economics. Northampton, Mass.: Edward Elgar.

Goetz, S. J., and D. L. Debrtin. In press. Why farmers quit: A county level analysis. American Journal of Agricultural Economics.

Green, G. P. 1995. Technology Adoption and Water Management in Irrigated Agriculture. Ph.D. Dissertation. Department of Agricultural and Resource Economics. University of California, Berkeley.

Gross, D., and the Editors of Forbes Magazine. 1996. Cyrus McCormick's reaper and the industrialization of farming. Pp. 22–39 in Forbes' Greatest Business Stories of All Time. New York: Wiley.

Harding, T. S. 1940. Science and Agricultural Policy. Farmers in a Changing World. Yearbook of Agriculture, 1940. Washington, D.C.: U.S. Department of Agriculture.

Hargrove, T. R. 1973. Agricultural Research: Impact on Swine. Special Report 72. Ames, Iowa: Agriculture and Home Economics Experiment Station, Iowa State University.

Hayami, Y., and V. M. Ruttan. 1985. Agricultural Development: An International Perspective. Baltimore: Johns Hopkins University Press.

Heinecke, C. 1994. African American migration and mechanized cotton harvesting, 1950–1960. Explorations in Economic History 31(4): 501–520.

Hightower, J. 1973. Hard Tomatoes, Hard Times. Cambridge, Mass: Schenkman.

Hoppe, R. A., and A. B. W. Effland. 1998. Minority and Women Farmers in the U.S. Agricultural Outlook, May. Washington, D.C.: Economic Research Service, U.S. Department of Agriculture.

Hoppe, R. A., R. Green, D. Banker, J. Z. Kalbacher, and S. E. Bentley. 1996. Structural and Financial Characteristics of U.S. Farms, 1993: 18th Annual Family Farm Report to Congress. Washington, D.C.: Economic Research Service, U.S. Department of Agriculture.

Hoppe, R. A., J. E. Perry, and D. Banker. 2000. ERS Farm Typology for a Diverse Agricultural Sector. Agriculture Information Bulletin No. 759 (AIB-759). Washington, D.C.: Resource Economics Division, Economic Research Service, U.S. Department of Agriculture.

Hubbell, B. J., M. C. Marra, and G. A. Carlson. 2000. Estimating the demand for a new technology: *Bt* cotton and insecticide policies. American Journal of Agricultural Economics 82(1): 118–132.

Huffman, W. E. 1974. Decision making: The role of education. American Journal of Agricultural Economics 56(1): 85–97.

REFERENCES

Huffman, W. E. 2000. Human capital, education, and agriculture. Staff paper series No. 338. Iowa State University, Department of Economics. Ames, Iowa.
Huffman, W. E., and R. E. Evenson. 1993. Science for Agriculture: A Long Term Perspective. Ames, Iowa: Iowa State University Press.
Huffman, W. E., and R. E. Evenson. 2001. Structural and productivity change in U.S. agriculture, 1950–1982. Agricultural Economics 24(2): 127–147.
Huffman, W. E., and J. Miranowski. 1981. An economic analysis of expenditures on agricultural experiment station research. American Journal of Agricultural Economics 63: 104–118.
Ilic, P. 1992. Southeast Asian small farm survey. University of California Cooperative Extension.
Iowa State University. 2000. Iowa Farms Community Supported Agriculture: 2000 Statewide List of Iowa CSA Farms, Producers, and Organizers. Extension Bulletin PM 1693. Revised July 2000. Ames, Iowa: Iowa State University.
Jensen, H. R. 1977. Farm management and production economics, 1940–1970. Pp. 3–89 in A Survey of Agricultural Economics Literature, Vol. 1. L. R. Martin, ed. Minneapolis, Minn.: University of Minnesota Press.
Just, R. E., and D. Zilberman. 1988. The effects of agricultural development policies on income distribution and technological change in agriculture. Journal of Development Economics 28: 192–216.
Kan-Rice, P. 1999. Radio reaches Hmong Farmers. California Agriculture 53(6): 14–15.
Kellogg Commission on the Future of State and Land-Grant Universities. 1999. Returning to Our Roots: The Engaged Institution. Washington, D.C.: National Association of State Universities and Land-Grant Colleges.
Khu, K. M., and E. P. Durrenberger, eds. 1998. Pigs, Profits, and Rural Communities. Albany, NY: State University of New York Press.
Kislev, Y., and W. Peterson. 1996. Economies of scale in agriculture: A reexamination of the evidence. Pp. 156–170 in Papers in Honor of D. Gale Johnson, J. M. Antle and D. A Sumner, eds. Chicago: University of Chicago Press.
Kislev, Y., and W. Peterson. 1981. Induced innovations and farm mechanization. American Journal of Agricultural Economics 63(3): 562–565.
Kislev, Y., and W. Peterson. 1982. Prices, technology, and farm size. Journal of Political Economy 90(3): 578–595.
Kitchen, N. R., K. A. Sudduth, S. J. Birrell, and S. C. Borgelt. 1996. Missouri Precision Agriculture Research and Education. Pp. 1091–1099 in Proceedings of the Third International Conference on Precision Agriculture. Jun 23–24. Minneapolis, Minn.
Koo, S., and J. R. Williams. 1996. Soil-Specific Production Strategies and Agricultural Contamination Levels in Northeast Kansas. Pp. 1079–1089 in Proceedings of the Third International Conference on Precision Agriculture. Jun 23–24. Minneapolis, Minn.
Lee, J. E., Jr. 1980. A framework for food and agricultural policy in the 1980s. Southern Journal of Agricultural Economics 12(1): 1–10.
Lewis, W. J, J. C. van Lenteren, S. C. Phatak, and J. H. Tumlinson III. 1997. A total system approach to sustainable pest management. Proceedings of the National Academy of Sciences USA 94(23): 12243–12248.
Lichtenberg, E. 1989. Land quality, irrigation development, and cropping patterns in the northern High Plains. American Journal of Agricultural Economics 71(1): 187–194.
Lichtenberg, E. D., D. Parker, and D. Zilberman. 1988. Marginal analysis of welfare costs of environmental policies: The case of pesticide regulation. American Journal of Agricultural Economics 70(4): 867–874.

Lipton, M., and R. Longhurst. 1989. New Seeds and Poor People. Baltimore: Johns Hopkins University Press.
Luft, L. D. 2000. Survey of Structure of Cooperative Extension. Moscow, Id.: University of Idaho Cooperative Extension System.
MacDonald, J. M., M. E. Ollinger, K. E. Nelson, and C. R. Handy. 1999. Consolidation in U.S. Meatpacking. Agricultural Economics Report No. 785. Washington, D.C.: Economic Research Service, U.S. Department of Agriculture. [Online] http://www.ers.usda.gov/publications/aer785/index.htm.
Manchester, A. C., and D. P. Blayney. 1997. The Structure of Dairy Markets: Past, Present, Future. Agricultural Economics Report 757. Washington, D.C.: Economic Research Service, U.S. Department of Agriculture.
Mann, C. K. 1978. Packages of Practices: A step at a time with clusters. Studies in Development 21: 73–81.
Marcus, A. I. 1985. Agricultural Science and the Quest for Legitimacy. Ames, Iowa: Iowa State University Press.
Marra, M. C., and G. A. Carlson. 1990. The decision to double crop: An application of expected utility theory using Stein's theorem. American Journal of Agricultural Economics 72(2): 337–345.
Martin, P. L. 1985. Labor in California Agriculture. Pp. 9–18 in Migrant Labor in Agriculture: An International Comparison, P. Martin, ed. Berkeley, Cali.: Giannini Foundation of Agricultural Economics.
Martin, P. L., and J. Perloff. 1997. Hired farm labor, ed. J. B. Siebert. Pp. 151–175 in California Agriculture: Issues and Challenges. Oakland, Cali.: Giannini Foundation of Agricultural Economics.
Martinez, S. 1999. Vertical Coordination in the Pork and Broiler Industries. Agricultural Economics Report No. 777. Washington, D.C.: Economic Research Service, U.S. Department of Agriculture.
McBride, W. D. 1997. Change in U.S. Livestock Production, 1969–1992. Agricultural Economics Report No. 754. Washington, D.C.: U.S. Department of Agriculture, Economic Research Service.
McWilliams, B., and D. Zilberman. 1996. Time of technology adoption and learning by doing. Economics of Innovation and New Technology 4: 131–154.
Millock, K., D. Sunding, and D. Zilberman. In press. Regulation of pollution with endogenous monitoring. Journal of Environmental Economics and Management. San Diego, Cali.: Academic.
Mountjoy, D. C. 2001. Ethnicity, multiple communities, and the promotion of conservation: Strawberries in California. In Interactions Among Agroecosystems and Human Rural Communities. C. B. Flora, ed. Boca Raton, FL.: CRC Press.
National Agricultural Research, Extension, Education, and Economics Advisory Board. 2000a. Recommendations for the Advisory Board: REE relevance and adequacy of funding review. Feb 18, 2000 Meeting. Chicago, IL. [Online] http://www.reeusda.gov/ree/advisory/reports/chicago.html.
National Agricultural Research, Extension, Education, and Economics Advisory Board. 2000b. Research and Education Recommendations for Small Farms. [Online] http://www.reeusda.gov/ree/advisory/reports/smallfarms.html.
NRC (National Research Council). 1989. Investing in Research: A Proposal to Strengthen the Agricultural, Food, and Environmental System. Washington, D.C.: National Academy Press.
NRC (National Research Council). 1994. Investing in the National Research Initiative: An Update of the Competitive Grants Program in the U.S. Department of Agriculture. Washington, D.C.: National Academy Press.

REFERENCES

NRC (National Research Council). 1995. Colleges of Agriculture at the Land Grant Universities: A Profile. Washington, D.C.: National Academy Press.
NRC (National Research Council). 1996. Colleges of Agriculture at the Land Grant Universities: Public Service and Public Policy. Washington, D.C.: National Academy Press.
NRC (National Research Council). 1997a. Intellectual Property Rights and Plant Biotechnology. Washington, D.C.: National Academy Press.
NRC (National Research Council). 1997b. Precision Agriculture in the 21st Century: Geospatial and Information Technologies in Crop Management. Washington, D.C.: National Academy Press.
NRC (National Research Council). 1999. Sowing Seeds of Change: Informing Public Policy in the Economic Research Service of USDA. Washington, D.C.: National Academy Press.
NRC (National Research Council). 2000a. National Research Initiative: A Vital Competitive Grants Program in Food, Fiber, and Natural-Resources Research. Washington, D.C.: National Academy Press.
NRC (National Research Council). 2000b. The Future Role of Pesticides in U.S. Agriculture. Washington, D.C.: National Academy Press.
OTA (U.S. Congress, Office of Technology Assessment). 1986. Technology, Public Policy, and the Changing Structure of American Agriculture. OTA-F-285. March. Washington, D.C.: Government Printing Office.
Olson, K. 1998. Precision Agriculture: Current Economic and Environmental Issues. Minneapolis, Minn.: Sixth Joint Conference on Food, Agriculture, and Environment. Aug. 31–Sept. 2.
Olmstead, A.L., and P. Rhode. 1993. Induced innovation in American agriculture: A reconsideration. Journal of Political Economy 101(1): 100–118.
Ortmann, G., G. F. Patrick, W. N. Musser, and D. H. Doster. 1993. Use of private consultants and other sources of information by large cornbelt farmers. Agribusiness: An International Journal 9(4): 391–402.
Ostrom, M. R., and F. H. Buttel. 1999. In Their Own Words: Wisconsin Farmers Talk About Dairying in the 1990s. Program on Agricultural Technology Studies Research Report No. 3. Madison, Wis.: College of Agricultural and Life Sciences, University of Wisconsin-Madison Cooperative Extension, University of Wisconsin-Extension.
Ostrom, M. R., and D. Jackson-Smith. 2000. The Use and Performance of Management Intensive Rotational Grazing Among Wisconsin Dairy Farms in the 1990s. PATS Research Report No. 8. Madison, Wis.: Program on Agricultural Technology Studies, University of Wisconsin-Madison.
Ostrom, M., D. Jackson-Smith, and S. Moon. 2000. Wisconsin dairy farmer views on university research and extension programs. Wisconsin Farm Research Summary. Summaries of Research from the Program on Agricultural Technology Studies 2(January): 1–6.
Pardey, P. G. 1992. A yardstick for international comparisons: An application to national agricultural research expenditures. Economic Development and Cultural Change 40(2): 333–349.
Parker, D. D., and D. Zilberman. 1996. The use of information services: The case of CIMIS. Agribusiness 12(3): 209–218.
Pearse, A. 1980. Seeds of Plenty, Seeds of Want. New York: Oxford University Press.
Perry, J., and D. Banker. 2000. Contracting changes how farms do business. Rural Conditions and Trends. Washington, D.C.: U.S. Department of Agriculture Economic Research Service 10(2): 50–55.

Peters, T. 1992. Liberation Management: Necessary Disorganization for the Nanosecond Nineties. New York: Knopf.
Pigford v. Glickman. 1997. U.S. 97-1978.
Porter, M. 1998. The Competitive Advantage of Nations. New York: Free Press.
Practical Farmers of Iowa. 2000. Local Food Brokering Project: 2000 Season Statistical Highlights.
Pray, C. E. 1993. Trends in food and agricultural R&D: Signs of declining competitiveness? Pp. 51–65 in U.S. Agricultural Research: Challenges and Options, R. Weaver, ed. Bethesda, Md.: Agricultural Research Institute.
Pretty, J. N. 1995. Regenerating Agriculture: Policies and Practice for Sustainable Growth and Self Reliance. Washington, D.C.: Joseph Henry Press.
Putler, D. S., and D. Zilberman. 1988. Computer use in agriculture: Evidence from Tulare County, California. American Journal of Agricultural Economics 70(4): 790–802.
Raffensperger, C., S. Peters, F. Kirschenmann, T. Schettler, K. Barrett, M. Hendrickson, D. Jackson, R. Voland, K. Leval, and D. Butcher. 1999. Defining Public Interest Research. Loka Alert 6: 3. The Loka Institute. [Online] http://www.loka.org/alerts/loka.6.3.htm.
Rehber, E. 1998. Vertical Integration in Agriculture and Contract Farming. Working Paper No. 46, May. Regional Hatch Project NE-165: Private Strategies, Public Policies, and Food System Performance. Storrs, Conn.: Food Marketing Policy Center, University of Connecticut. [Online] http://agecon.lib.umn.edu/cgi-bin/pdf_view.pl?paperid=882.
Rhoades, R., and R. Booth. 1982. Farmer back to farmer: A model for generating acceptable technology. Agricultural Administration 11: 127–137.
Richardson, J. G., J. Knight, F. May, and M. McAlister. 1998. Assessment of program delivery to small farmers: Fax information center system at satellite locations. Journal of Applied Communications 82(1): 21–30.
Rogers, E. 1995. Diffusion of Innovations. Fourth Edition. New York: Free Press.
Rominger, R. 1999. Address to the 19th Annual Ecological Farming Conference, Pacific Grove, Cali., Jan. 21, 1999. Reprinted as Guest Commentary in the Information Bulletin of the Organic Farming Research Foundation. Summer(6): 5.
Rothschild, M. F., and L. L. Christian. 1988. Genetic control of front leg weakness in Duroc swine. I. Direct response to five generations of divergent selection. Livestock Production Science 19: 459–471.
Ruttan, V. 2000. Technology, Growth, and Development: An Induced Innovation Perspective. Oxford, U.K.: Oxford University Press.
Salant, P., M. Smale, and W. Saupe. 1986. Farm Viability: Results of the USDA Family Farm Surveys. RDRR-60. Washington, D.C.: Economic Research Service, U.S. Department of Agriculture.
Sawyer, J. E. 1994. Concepts of variable rate technology with considerations for fertilizer application. Journal of Production Agriculture 7(2): 195–201.
Schmitz, A., and D. Seckler. 1970. Mechanized agriculture and social welfare: The case of the tomato harvester. American Journal of Agricultural Economics 52(4): 569–577.
Schultz, T. W. 1964. Transforming Traditional Agriculture. New Haven, Connecticut: Yale University Press.
Schultz, T. W. 1975. The value of the ability to deal with disequilibria. Journal of Political Economy 13: 827–846.
Sisson, C. 1982. Tax Burdens in American Agriculture. Ames, Iowa: Iowa State University Press.

REFERENCES

Sommer, J. E., R. A. Hoppe, R. C. Green, and P. J. Korb. 1998. Structural and Financial Characteristics of U.S. Farms, 1995: 20th Annual Family Farm Report to the Congress. Agriculture Information Bulletin No. 746 (AIB-746). Washington, D.C.: Resource Economics Division, Economic Research Service, U.S. Department of Agriculture.

Stanton, B. F. 1993. Changes in farm size and structure in American agriculture in the twentieth century. Pp. 42–70 in Size, Structure, and the Changing Face of American Agriculture. A. Hallam, ed. Boulder, Colo: Westview Press.

Sunding, D., and D. Zilberman. 2001. The agricultural innovation process: Research and technology adoption in a changing agricultural sector. Pp. 1–103 in The Handbook of Agricultural Economics, G. C. Rausser and B. Gardner, eds. Amsterdam: North-Holland.

Swinton, S., and J. Lowenberg-DeBoer. 1998. Evaluating the profitability of site-specific farming. Journal of Production Agriculture 11(4): 439–446.

Thirtle, C. G., and V. W. Ruttan. 1987. The Role of Demand and Supply in the Generation and Diffusion of Technical Change. New York: Harwood Academic.

Thompson, J., and I. Guijt. 1999. Sustainability indicators for analysing the impacts of participatory watershed management programmes. Pp. 13–26 in Fertile Ground: The Impacts of Participatory Watershed Management, F. Hinchcliffe, J. Thompson, J. Pretty, I. Guijt, and P. Shah, eds. London: Intermediate Technology Publications, Ltd.

Tweeten, L. 1989. P. 112 in Farm Policy Analysis. Boulder, Colo.: Westview Press.

Tweeten, L. 1993. Government commodity program impacts on farm numbers. Ch. 13 in Size, Structure, and the Changing Face of American Agriculture, A. Hallam, ed. Boulder, Colo.: Westview Press.

Tweeten, L. G., and C. B. Flora. 2001. Vertical Coordination of Agriculture in Farming-Dependent Areas. Task Force Report No. 137. Ames, Iowa: Council for Agricultural Science and Technology.

Tyner, W. E., and M. Boehlje. 1997. Food System 21: Gearing Up for the New Millennium. EC-710, Purdue University Cooperative Extension Service, Department of Agricultural Economics, and the faculty of the School of Agriculture. West Lafayette, IN.

U.S. Bureau of the Census. 1980. Statistical Abstract of the United States: 1980 (101st Edition). Washington, D.C.: U.S. Department of Commerce.

U.S. Bureau of the Census. 1900–1992. Censuses of Agriculture. Washington, D.C.: U.S. Department of Commerce.

U.S. Bureau of the Census. 1930–1992. Censuses of Agriculture. Washington, D.C.: United States Department of Commerce.

U.S. Congress. 1862. P.L. (Public Law) 301 et seq., 12 Stat. 503. Morrill Act of 1862.

U.S. Congress. 1887. P.L. (Public Law) 361a et seq., 24 Stat. 440. Hatch Act of 1887.

U.S. Congress. 1914. P.L. (Public Law) 341 et seq., 38 Stat. 372. Smith-Lever Act of 1914.

U.S. Congress. 1970. P.L. (Public Law) 2321 et seq. Plant Variety Protection Act of 1970.

U.S. Congress. 1980. P.L. (Public Law) 96-517. Government Patent Policy Act of 1980.

U.S. Congress. 1986. P.L. (Public Law) 99-502. Federal Technology Transfer Act of 1986.

U.S. Congress. 1992. P.L. (Public Law) 102-325. Higher Education Reauthorization Act of 1992.

U.S. Congress. 1994. P.L. (Public Law) 103-382. Equity in Educational Land Grant Status Act of 1994.

U.S. Congress. 1996. P.L. (Public Law) 104-127. Federal Agriculture Improvement and Reform Act of 1996.

U.S. Congress, 1998. P.L. (Public Law) 105-85. Agricultural Research, Extension, and Education Reform Act of 1998.
USDA (U.S. Department of Agriculture). 1981. A Time to Choose: Summary Report on the Structure of Agriculture. Washington, D.C.: U.S. Department of Agriculture.
USDA (U.S. Department of Agriculture). 1986. Selected Current Research Information System (CRIS) Funding Summaries, FY 1986. Unpublished tables. Washington, D.C.: U.S. Department of Agriculture.
USDA (U.S. Department of Agriculture). 1996. Charting the Course for the Cooperative Extension System Federal Agenda: The Working Group Report for Dr. Karl Stauber, Undersecretary, Research, Education, and Economics, Washington, D.C.: U.S. Department of Agriculture.
USDA (U.S. Department of Agriculture). 1997a. Civil Rights at the U.S. Department of Agriculture: A Report of the Civil Rights Action Team. [Online] http://www.usda.gov/news/civil/cr_index.htm.
USDA (U.S. Department of Agriculture). 1997b. Selected Current Research Information System (CRIS) Funding Summaries, FY 1997. [Online] http://cris.csrees.usda.gov/star/crisfin.htm. Washington, D.C.: U.S. Department of Agriculture.
USDA (U.S. Department of Agriculture). 1997c. Packers and Stockyards Statistical Report, 1995 Reporting Year, SR-97-1. P. 49 in Grain Inspection Packers and Stockyards Administration. Washington, D.C.: U.S. Department of Agriculture.
USDA (U.S. Department of Agriculture). 1998a. 1998 Agricultural Resource Management Study. Economic Research Service, U.S. Department of Agriculture.
USDA (U.S. Department of Agriculture). 1998b. A Time to Act: A Report of the U.S. National Commission on Small Farms. Washington, D.C.: U.S. Department of Agriculture.
USDA (U.S. Department of Agriculture). 1998c. Direct Marketing Action Plan. Washington, D.C.: Agricultural Marketing Service, U.S. Department of Agriculture. [Online] http://www.ams.usda.gov/directmarketing/frmplan.htm.
USDA (U.S. Department of Agriculture). 1998d. Ten Years of SARE: A Decade of Programs, Partnerships, and Progress in Sustainable Agriculture Research and Education. Washington, D.C.: Cooperative State Research, Education, and Extension Service, U.S. Department of Agriculture.
USDA (U.S. Department of Agriculture). 1999a. Agricultural Resource Management Study. Washington, D.C.: Economic Research Service, U.S. Department of Agriculture.
USDA (U.S. Department of Agriculture). 1999b. Agricultural Research. 47(10).
USDA (U.S. Department of Agriculture). 1999c. 1997 Census of Agriculture. Washington, D.C.: National Agricultural Statistical Service, U.S. Department of Agriculture.
USDA (U.S. Department of Agriculture). 1999d. Glickman announces reduction in Salmonella in raw meat; provides $9.6 million for food safety grants. USDA Press Release No. 0399.99. Oct. 7, 1999. Washington, D.C.: U.S. Department of Agriculture. [Online] http://www.usda.gov/news/releases/1999/10/0399.
USDA (U.S. Department of Agriculture). 1999e. Report of the Strategic Task Force on USDA Research Facilities, Washington, D.C.: U.S. Department of Agriculture.
USDA (U.S. Department of Agriculture). 1999f. Selected Current Research Information System (CRIS) Funding Summaries, FY 1999. [Online] http://cris.csrees.usda.gov/star/crisfin.htm. Washington, D.C.: U.S. Department of Agriculture.

REFERENCES

USDA (U.S. Department of Agriculture). 2000a. Agricultural Resources and Environmental Indicators, 2000. Washington, D.C.: Economic Research Service, Resource Economics Division, U.S. Department of Agriculture. [Online] http://www.ers.usda.gov/Emphases/Harmony/issues/arei2000/arei2000.htm.

USDA (U.S. Department of Agriculture). 2000b. Contribution of ARS Research to Small Farms. April 18, 2000. Washington, D.C.: Agricultural Research Service, U.S. Department of Agriculture.

USDA (U.S. Department of Agriculture). 2000c. National Research Initiative Competitive Grants Program 2001 Program Description. Washington, D.C.: Cooperative State Research, Extension, and Education Service, U.S. Department of Agriculture. [Online] http://www.reeusda.gov/nri/programs/progdesc/nridtop.htm.

USDA (U.S. Department of Agriculture). 2000d. Preliminary Report and Summary, Initiative for Future Agriculture and Food Systems. Washington, D.C.: Cooperative State Research, Education, and Extension Service, U.S. Department of Agriculture.

USDA (U.S. Department of Agriculture). 2000e. Sustainable Agriculture Research and Education Program. North Central Regional Technical Committee. Washington, D.C.: Cooperative State Research, Education, and Extension Service, U.S. Department of Agriculture. [Online] http://www.sare.org/ncrsare/leaders.htm.

USDA (U.S. Department of Agriculture). 2001a. Federal State Marketing Improvement Program, Fiscal Year 2001 Priorities. Washington, D.C.: Agricultural Marketing Service, U.S. Department of Agriculture. [Online] http://www.ams.usda.gov/tmd/fsmip.htm.

USDA (U.S. Department of Agriculture). 2001b. Professional Employment by Series. Human Resources Division of Research, Education, and Economics. Washington, D.C.: Agricultural Research Service, U.S. Department of Agriculture. 10 June.

USDA (U.S. Department of Agriculture). 2001c. Research Emphasis: Harmony Between Agriculture and the Environment: Current Issues. Washington, D.C.: Economic Research Service, U.S. Department of Agriculture. [Online] http://www.ers.usda.gov/Emphases/Harmony/issues/organic/organic.html#datatables.

USDA (U.S. Department of Agriculture). 2001d. Research Scientist Workforce. Human Resources Division of Research, Education, and Economics. Washington, D.C.: Agricultural Research Service, U.S. Department of Agriculture. 10 June.

USDA (U.S. Department of Agriculture). 2001e. South Atlantic Sea, Small Farm Survival Project for the Southern Coastal Plain. Washington, D.C.: Agricultural Research Service, U.S. Department of Agriculture.

USDA (U.S. Department of Agriculture). 2001f. Structural and financial characteristics of U.S. Farms: 2001 family farm report. R. A. Hoppe, ed. Washington, D.C.: Resource Economics Division, Economic Research Service, U.S. Department of Agriculture. Agriculture Information Bulletin No. 768. [Online] http://www.ers.usda.gov/Publications/aib768/.

U.S. General Accounting Office (GAO). 2001. Farm programs: Information on recipients of federal payments. Report to the Chairman, Committee on Agriculture, Nutrition, and Forestry, U.S. Senate. Report Number GAO-01-606.

U.S. Patent and Trademark Office (PTO). 1975–1998. Statistical Reports Available for Viewing. Washington, D.C.: Technology Assessment and Forecast Branch, U.S. Patent and Trademark Office.
[Online] http://www.uspto.gov/web/offices/ac/ido/oeip/taf/reports.htm.

University of California. 1998. Swiss pharmaceutical company Novartis commits $25 million to support biotechnology research at UC Berkeley. Press Release. November 23. [Online] http://navigation.helper.realnames.com/framer/1/112/default.asp?realname=UC+Berkeley&url=http%3A%2F%2Fwww%2Eberkeley%2Eedu%2F&frameid=1&providerid=112&uid=30004204.

University of California. 1999. Small farms: Stories of success and struggle. California Agriculture 53(6).

WARDA (West Africa Rice Development Association). 2000. Participatory Varietal Selection: The Flame Spreads into 2000. Proceedings of the Participatory Rice Improvement and Gender/User Analysis Workshop (PRIGA). 17–21 April. Bouaké, Côte d'Ivoire: West Africa Development Association.

Warner, P. D., R. Rennekamp, and M. Nall. 1996. Structure and Function of the Cooperative Extension Service. Kentucky Cooperative Extension Service. Lexington, Ky.: University of Kentucky.

Watkins, K. B., Y. C. Lu, and W. Y Huang. 1998. Economic returns and environmental impacts of variable nitrogen fertilizer and water applications. Proceedings of the Fourth International Conference on Precision Agriculture. St. Paul, Minn.: July 19–22.

Weber, M. 1978. Economy and Society. Berkeley, Cali.: University of California Press.

Weibers, U. C. 1992. Economic and environmental effects of pest management information and pesticides: The case of processing tomatoes in California, Ph.D. Dissertation. Berlin: Department of Agricultural Sciences. University of Berlin.

Williams-Derry, C., and K. Cook. 2000. Green Acre$: How Taxpayers are Subsidizing the Demise of the Family Farm. Washington, D.C.: Environmental Working Group. [Online] http://www.ewg.org/pub/home/reports/greenacres/exec.html.

Wolf, S. A., ed. 1998. Privatization of Information and Agricultural Industrialization. Boca Raton, Fla.: CRC Press.

World Bank Group. 1999. Entering the 21st Century: World Development Report 1999/2000. [Online] http://www.worldbank.org/wdr/2000/fullreport.html.

Wozniak, G. D. 1993. Joint information acquisition and new technology adoption: Late versus early adoption. The Review of Economics and Statistics 75(3): 438–445.

Wright, B. 1998. Public germplasm development at a crossroads: Biotechnology and intellectual property. California Agriculture 52: 8–13.

Zilberman, D., A. Schmitz, G. Casterline, E. Lichtenberg, and J. B. Siebert. 1991. The economics of pesticide use and regulation. Science 253(5019): 518–522.

Zilberman, D. C., Yarkin, and A. Heiman. 1999. Agricultural biotechnology: Economic and international implications. Pp. 144–161 in Food Security, Diversification, and Resource Management: Refocusing the Role of Agriculture? G. H. Peters and J. von Braun, eds. Aldershot, England: Ashgate.

Appendixes

Appendix A

Committee to Review the Role of Publicly Funded Agricultural Research on the Structure of U.S. Agriculture

Public Workshop

NATIONAL RESEARCH COUNCIL
BOARD ON AGRICULTURE AND NATURAL RESOURCES

University of Wisconsin, Madison

November 19, 1999

AGENDA

8:30 – 9:00 am	**Welcome and Introductions** Anthony Earl, Committee Chair; Lee Paulson, Project Director
8:45	**Agricultural Research Service Overview** Edward Knipling, Associate Adminstrator
9:30	**Competitive Research Grants and Awards Management Overview**

	Sarah (Sally) Rockey, Deputy Administrator, Cooperative State Research, Education, and Extension Service, U.S. Department of Agriculture
10:15	Break
10:30	**Effects of Publicly Funded Agricultural Research on Structure** Robert Evenson, Professor of Economics, Yale University
11:15	**Producer Perspective** Mike Wehler, Upland Prairie Farms, Wisconsin
12:00 noon	Lunch
1:00	**Producer Perspective** Marlyn Jorgensen, Jorg-Anna Farms, Garrison, Iowa
1:45	**Federal/Producer Interactions and Perspectives** Roger Gerrits, Agricultural Consultant (formerly Agricultural Research Service, U.S. Department of Agriculture)
2:30	**Constraints and Obligations to Providing Publicly Funded Research to Underserved Citizens for the Broad Public Good** Margaret Krome, Michael Fields Agricultural Institute, East Troy, Wisconsin
3:15	Break
3:30	**Swine Production Systems** Jay Harmon, Associate Professor and Extension Agricultural Engineer, Iowa State University
4:15	**Science, Technology, and the Structure of the Dairy Farm Sector: Review of Research Results by the Program on Agricultural Technology Studies, University of Madison, Wisconsin** Douglas Jackson-Smith, Co-Director
5:00	**Producer Perspective**

APPENDIX A

James Van Der Pol, Minnesota Institute for
Sustainable Agriculture

5:45   Adjourn

Appendix B

Committee to Review the Role of Publicly Funded Agricultural Research on the Structure of U.S. Agriculture

Public Workshop

NATIONAL RESEARCH COUNCIL
BOARD ON AGRICULTURE AND NATURAL RESOURCES

Arnold and Mabel Beckman Center

January 18, 2000

AGENDA
8:00 am   **Welcome and Introductions**
                 Anthony Earl, Committee Chair

8:10   **Sustainable Agriculture and Small-Scale Linkages**
                 Gail Feenstra, University of California, Davis

8:40   **Small Farms Commission and Minority Perspectives**
                 Desmond Jolly, University of California, Davis

9:10   **Capturing Value of Publicly Funded Research and the Structure of Agriculture**

	Chuck Hassebrook, Center for Rural Affairs
9:40	**Q&A Session**
10:10 – 10:30	Break
10:30	**Overall Funding for Agricultural Research and Information Transfer Issues** Noel Keen, University of California, Riverside
11:00	**Impact of Biotechnology Research and Producer Access to Information** Ken Olson, American Farm Bureau
11:30	**Genomics and GMOs: Dealing with Public Opinion and Policy Needs from an International Perspective** Robert Goodman, University of Wisconsin, Madison
12:00 noon	**Q&A Session**
12:30 – 1:30 pm	Lunch
1:30	**Small Farms and Federal Funding—Farmer's Perspective** Glenn Anderson, Organic Farms, Hilmar, California
2:00	**Minorities and Public Research** Daniel Mountjoy, Natural Resources Conservation Service, U.S. Department of Agriculture
2:30	**Sustainable Agriculture and the Salad Bar Project** Michelle Mascarenhas, Occidental College, Los Angeles
3:00	**Q&A Session**
3:30 – 3:50	Break
3:50	**Publicly Funded Agricultural Research— Perspective from the Biological Sciences** Cal Qualset, University of California, Davis

4:20	**Creating a Small Farm Research and Education Program in a Traditional Context** James Zuiches, Washington State University
4:50	**Role of ESCOP and Linkages**
5:20	**Q&A Session**
5:50	**Closing Remarks** Anthony S. Earl, Committee Chair
6:00	Adjourn

Appendix C

Table C.1 U.S. public (USDA and State Agricultural Experiment Stations [SAES]) and private agricultural research funds by performing organization, 1888–1990 (millions of 1984 dollars)

Year	Price Index for Agricultural Research (1984–1.0)	Public Agricultural Research			Private Agricultural Research [a]
		USDA	SAES	Total	
1888	0.0472	3.093	15.254	18.347	
1889	0.0472	3.030	15.254	18.284	
1890	0.0472	4.767	19.513	24.280	
1891	0.0469	4.435	19.446	23.881	
1892	0.0458	4.214	22.620	26.834	
1893	0.0471	4.119	20.658	24.777	
1894	0.0444	4.392	23.086	27.477	
1895	0.0454	5.374	24.361	29.736	32.400
1896	0.0452	4.469	26.394	30.863	
1897	0.0452	4.558	26.372	30.929	
1898	0.0468	4.423	27.201	31.624	
1899	0.0493	5.051	24.665	29.716	

(continued)

Year	Price Index for Agricultural Research (1984–1.0)	Public Agricultural Research			Private Agricultural Research a/
		USDA	SAES	Total	
1900	0.0521	5.067	24.434	29.501	
1901	0.0522	8.927	26.284	35.211	
1902	0.0547	11.865	27.148	39.013	
1903	0.0571	12.907	28.161	41.068	
1904	0.0557	14.794	30.844	45.637	
1905	0.0567	14.797	30.459	45.256	53.100
1906	0.0583	18.971	41.252	60.223	
1907	0.0605	25.256	45.934	71.190	
1908	0.0596	26.510	55.084	81.594	
1909	0.0617	39.287	54.587	93.874	
1910	0.0632	35.570	61.487	97.057	
1911	0.0612	45.033	65.523	110.556	
1912	0.0640	50.016	70.188	120.203	
1913	0.0664	47.666	76.596	124.262	
1914	0.0655	63.771	86.107	149.878	
1915	0.0668	60.419	86.482	146.901	101.600
1916	0.0753	65.511	75.219	140.730	
1917	0.0933	59.893	64.652	124.544	
1918	0.1020	64.490	65.147	129.637	
1919	0.1091	74.601	65.060	139.661	
1920	0.1225	63.184	66.947	130.131	
1921	0.1062	149.369	81.742	231.111	
1923	0.1087	148.289	92.355	240.644	
1924	0.1089	151.598	96.814	248.411	
1925	0.1130	195.858	96.664	292.522	140.000
1926	0.1133	206.346	110.477	316.823	
1927	0.1126	190.107	119.547	309.654	
1928	0.1153	199.922	133.591	333.513	
1929	0.1158	250.924	144.940	395.864	
1930	0.1116	326.093	164.095	490.188	
1931	0.1064	328.769	173.571	502.340	
1932	0.1018	304.715	173.320	478.035	
1933	0.0993	289.940	159.980	449.919	
1934	0.1003	276.670	143.918	420.588	
1935	0.0992	285.121	153.972	439.093	405.400
1936	0.1012	281.611	164.526	446.136	
1937	0.1067	257.460	167.994	425.455	
1938	0.1028	277.986	195.039	473.025	
1939	0.1029	340.214	202.634	542.847	

APPENDIX C

(continued)

Year	Price Index for Agricultural Research (1984–1.0)	Public Agricultural Research			Private Agricultural Research a/
		USDA	SAES	Total	
1940	0.1035	318.406	207.362	525.768	
1941	0.1077	307.697	210.594	518.292	
1942	0.1134	296.984	202.019	499.004	
1943	0.1176	293.980	207.993	501.973	
1944	0.1252	250.072	217.236	467.308	
1945	0.1247	260.634	227.466	488.099	346.300
1946	0.1371	251.014	243.027	494.041	
1947	0.1584	358.794	261.521	620.316	
1948	0.1731	415.881	295.904	711.785	
1949	0.1711	282.548	331.473	614.021	
1950	0.1821	164.618	357.062	521.680	
1951	0.1985	157.149	352.932	510.081	
1952	0.2038	168.391	375.020	543.410	
1953	0.2106	160.095	385.408	545.503	
1954	0.2183	186.702	409.372	596.074	
1955	0.2263	188.785	435.024	623.809	
1956	0.2395	194.418	419.415	613.833	890.600
1957	0.2517	239.050	449.670	688.721	994.100
1958	0.2490	271.940	508.446	780.386	1189.900
1959	0.2640	279.091	511.504	790.595	1086.700
1960	0.2717	274.056	523.905	797.961	1175.300
1961	0.2788	294.756	540.219	834.975	1120.800
1962	0.2896	295.328	555.435	850.763	1159.400
1963	0.2970	309.165	581.313	890.478	1210.500
1964	0.3092	343.797	605.049	948.845	1258.600
1965	0.3236	384.778	631.100	1015.878	1367.800
1966	0.3416	376.259	661.212	1037.471	1431.800
1967	0.3570	372.756	691.476	1064.232	1476.200
1968	0.4049	319.331	624.194	943.524	1386.200
1969	0.3983	345.205	649.608	994.813	1517.100
1970	0.4183	350.210	673.653	1023.863	1486.000
1971	0.4328	365.016	692.740	1057.756	1479.500
1972	0.4519	459.533	765.751	1225.284	1494.400
1973	0.4837	445.979	795.127	1241.106	1579.700
1974	0.5286	424.410	801.901	1226.311	1602.700
1975	0.5654	440.021	852.821	1292.842	1577.800
1976	0.5921	607.782	1091.464	1699.247	1569.600
1977	0.6253	507.660	948.698	1456.359	1973.500
1978	0.6656	524.340	974.890	1499.231	2092.700
1979	0.7240	495.119	991.779	1486.898	2146.000

(continued)

Year	Price Index for Agricultural Research (1984–1.0)	Public Agricultural Research			Private Agricultural Research [a]
		USDA	SAES	Total	
1980	0.7484	510.750	1075.402	1586.152	2300.100
1981	0.8183	541.317	1091.845	1633.163	2311.300
1982	0.8716	508.608	1092.585	1601.193	2348.800
1983	0.9518	500.718	1046.763	1547.481	2380.800
1984	1.0000	482.492	1059.343	1541.835	2444.700
1985	1.0531	502.702	1088.175	1590.877	2550.100
1986	1.0944	471.241	1125.848	1597.089	2660.100
1987	1.1383	482.894	1141.861	1624.754	2774.800
1988	1.1210	521.967	1225.893	1747.860	2894.500
1989	1.2729	468.344	1170.289	1638.633	3019.300
1990	1.3379	458.988	1193.254	1652.242	3149.500
1991	1.376	483.127	1220.447	1703.574	3269.847
1992	1.411	498.681	1219.531	1718.212	3222.913
1993	1.444	493.319	1209.504	1702.823	3400.644
1994	1.488	490.571	1227.073	1717.644	3389.671
1995	1.532	485.758	1218.649	1704.402	3595.907
1996	1.580	457.604	1195.134	1652.738	3554.873
1997	1.612	462.983	1201.520	1664.503	3676.895

[a] Estimates of private agricultural research expenditures were derived as a decade average for the period 1890–1950, and for 1985 and later. The numbers for private agricultural research are also an estimate.

SOURCE: Updated from Huffman, W. E. and R. E. Evenson. 1993. Science for Agriculture: A Long Term Perspective. Ames, Iowa: Iowa State University Press, pp. 95–96.

Appendix D
Economic Research Service Farm Typology

The U.S. Department of Agriculture's Economic Research Service has developed a farm classification to divide U.S. farms into mutually exclusive and more homogeneous groups. The farm typology focuses on "family farms," or farms organized as proprietorships, partnerships, and family corporations that are not operated by a hired manager. To be complete, however, the typology also considers nonfamily farms.

Small Family Farms (annual sales less than $250,000)

Limited-resource farms. Any small farm with (1) gross sales of less than $100,000, (2) total farm assets of less than $150,000, and (3) total operator household income of less than $20,000. Limited-resource farmers report farming, a nonfarm occupation, or retirement as their major occupation.

Retirement farms. Small farms whose operators report they are retired. (Excludes limited-resource farms operated by retired farmers.)

Residential/lifestyle farms. Small farms whose operators report they had a major occupation other than farming. (Excludes limited-resource farms with operators reporting a nonfarm major occupation.)

Farming-occupation/low-sales. Small farms with annual sales of less than $100,000 whose operators report farming as their major occupation. (Excludes limited-resource farms whose operators report farming as their major occupation.)

Farming-occupation/high-sales. Small farms with annual sales between $100,000 and $249,999 whose operators report farming as their major occupation.

Other Farms

Large family farms. Annual sales between $250,000 and $499,999.

Very large family farms. Annual sales of $500,000 or more.

Nonfamily farms. Farms organized as nonfamily corporations or cooperatives, as well as farms operated by hired managers.

About the Authors

Anthony S. Earl, Chair, has been a partner at the Quarles and Brady Law Firm in Madison, Wisconsin, since 1987. He served as the 40th governor of the State of Wisconsin (1983–1986). Earl has extensive expertise in environmental law and policy, and he is involved in many civic activities. An advocate of environmental civic responsibility, as governor, Earl successfully advanced through the legislature a significant number of initiatives in the areas of education, equal opportunity, economic development, and protection of the environment. Earl served on the NRC Board on Agriculture and Natural Resources from 1996 to 1999. He chaired the NRC Committee on the Future of Colleges of Agriculture in the Land Grant University System. Earl received his B.A. from Michigan State University and a J.D. from the University of Chicago (1961).

Michael Boehlje is professor in the Department of Agricultural Economics at Purdue University in West Lafayette, Indiana. He has extensive expertise in farm and agribusiness management and finance. Boehlje is a former head of the Department of Agricultural and Applied Economics at the University of Minnesota. He also served as assistant dean for the College of Agriculture, Iowa State University and as assistant director of the Iowa Agricultural Experiment Station. Boehlje conducts research and teaches in the area of farm and agribusiness management and finance. His research interests include alternative systems of coordination in the food and industrial product chain, industrialization of agriculture, and alternative financial and organizational structures for farm and agribusiness firms. Boehlje's work focuses on strategic planning, visioning finance, and business policy. He received an M.S. in 1968 and Ph.D. in 1971, both in agricultural economics, from Purdue University.

R. Dean Boyd is director of nutrition at Pig Improvement Company (PIC) USA in Franklin, Kentucky. He has expertise in the dynamics of animal nutrition and experience with industrial and academic research management. Boyd manages nutrition-genotype research and provides nutrient recommendations and technical service to customers. He also manages the nutrition program for PIC farms and joint-venture partners, and is a member of the technical strategy team for PIC Group (UK). Before joining PIC, Boyd was professor of animal science at Cornell University. His research group made important contributions to explaining of the regulation of nutrient use for lean growth, methods to improve the efficiency of amino acid use, and biologic potential for growth. Boyd was a member of the NRC Subcommittee on Role of Metabolic Modifiers on Animal Nutrient Requirements. He received his Ph.D. in animal nutrition from the University of Nebraska in 1979.

Frederick H. Buttel is professor and chair of the Department of Rural Sociology and professor of environmental studies at the University of Wisconsin. Buttel also is associate director of the university's Program on Agricultural Technology Studies. He has expertise in rural and environmental sociology, and in the sociology of agrarian systems. Buttel's research interests include environmental sociology and policy, technology and social change, political sociology, sociology of development, theory of sociology, and sociology of science. He served as a member of the NRC committee that produced *Managing Global Genetic Resources* (1993). Buttel holds master's degrees in rural sociology, from University of Wisconsin, Madison (1972), and in forestry and environmental studies, from Yale University (1973). He received his Ph.D. in sociology from the University of Wisconsin, Madison in 1975.

Cornelia B. Flora is director of the North Central Regional Center for Rural Development and professor of sociology at Iowa State University. She has extensive background in rural sociology, agriculture, and in rural development. Her research interests include rural America and global restructuring, science and sustainability, and rural economic development through self-development strategies. Before Flora's appointment at Iowa State University, she was professor and head of the Department of Sociology at Virginia Polytechnic Institute & State University, university distinguished professor at Kansas State University, and program advisor for agricultural development at the Ford Foundation. Flora serves on the NRC Board on Agriculture and Natural

Resources. She received her M.S. in rural sociology in 1966, and a Ph.D. in development sociology in 1970, both from Cornell University.

Peter J. Goldmark is the owner and operator of a 7,000-acre farm (Double J. Ranch, Inc.) that is evenly split between farmland and pasture land. He also is the founder and chief scientist of a biotechnology research laboratory, DJR Research, Inc., in Okanogan, Washington. Goldmark has expertise and hands-on experience in farming and ranching, and experience with regulatory and policy issues. In 1993, he was the director of the Washington State Department of Agriculture, and he currently serves on the Washington State University Board of Regents. He also served as chair of the Governor's Council on Agriculture and the Environment, and he has held many positions within the Washington Association of Wheat Growers. Goldmark received his Ph.D. in molecular biology from the University of California, Berkeley in 1971.

Frederick Kirschenmann is director of the Leopold Center for Sustainable Agriculture in Ames, Iowa, and founder and president of Farm Verified Organic, Inc., in Medina, North Dakota, a private certification agency for organic farmers. He also is the manager of Kirschenmann Family Farms, a 3,500-acre grain and livestock operation, which he converted into an organic farm. Kirschenmann's expertise includes issues related to sustainable agriculture and farm operations. He is former dean and professor at Curry College in Boston, Massachusetts. Kirschenmann has been active in numerous sustainable and organic agriculture organizations. He has published numerous articles and book chapters on sustainable agriculture and related topics. Kirschenmann earned a Ph.D. in historical theology from the University of Chicago in 1964.

David Zilberman is professor and chair in the Department of Agricultural and Resource Economics and director of the Center for Sustainable Resource Development at the University of California, Berkeley. His expertise includes natural resource economics, agricultural research policy, and adoption of technologies at the farm level. He served on the NRC Committee on the Future Role of Pesticides in U.S. Agriculture. Zilberman received his B.A. in economics/statistics in 1971 from Tel Aviv University, Israel, and his Ph.D. in agricultural and resource economics in 1979 from the University of California, Berkeley.

Board on Agriculture and Natural Resources Publications

Policy and Resources

Agricultural Biotechnology: Strategies for National Competitiveness (1987)
Agriculture and the Undergraduate: Proceedings (1992)
Agriculture's Role in K-12 Education: A Forum on the National Science Education Standards (1998)
Alternative Agriculture (1989)
Brucellosis in the Greater Yellowstone Area (1998)
Colleges of Agriculture at the Land Grant Universities: Public Service and Public Policy (1996)
Colleges of Agriculture at the Land Grant Universities: A Profile (1995)
Designing an Agricultural Genome Program (1998)
Designing Foods: Animal Product Options in the Marketplace (1988)
Ecological Monitoring of Genetically Modified Crops (2001)
Ecologically Based Pest Management: New Solutions for a New Century (1996)
Ensuring Safe Food: From Production to Consumption (1998)
Environmental Effects of Transgenic Plants: The Scope and Adequacy of Regulation (2002)
Forested Landscapes in Perspective: Prospects and Opportunities for Sustainable Management of America's Nonfederal Forests (1997)
Future Role of Pesticides for U.S. Agriculture (2000)

Genetic Engineering of Plants: Agricultural Research Opportunities and Policy Concerns (1984)
Genetically Modified Pest-Protected Plants: Science and Regulation (2000)
Incorporating Science, Economics, and Sociology in Developing Sanitary and Phytosanitary Standards in International Trade (2000)
Investing in Research: A Proposal to Strengthen the Agricultural, Food, and Environmental System (1989)
Investing in the National Research Initiative: An Update of the Competitive Grants Program in the U.S. Department of Agriculture (1994)
Managing Global Genetic Resources: Agricultural Crop Issues and Policies (1993)
Managing Global Genetic Resources: Forest Trees (1991)
Managing Global Genetic Resources: Livestock (1993)
Managing Global Genetic Resources: The U.S. National Plant Germplasm System (1991)
National Research Initiative: A Vital Competitive Grants Program in Food, Fiber, and Natural-Resources Research (2000)
New Directions for Biosciences Research in Agriculture: High-Reward Opportunities (1985)
Pesticide Resistance: Strategies and Tactics for Management (1986)
Pesticides and Groundwater Quality: Issues and Problems in Four States (1986)
Pesticides in the Diets of Infants and Children (1993)
Precision Agriculture in the 21st Century: Geospatial and Information Technologies in Crop Management (1997)
Professional Societies and Ecologically Based Pest Management (2000)
Rangeland Health: New Methods to Classify, Inventory, and Monitor Rangelands (1994)
Regulating Pesticides in Food: The Delaney Paradox (1987)
Soil and Water Quality: An Agenda for Agriculture (1993)
Soil Conservation: Assessing the National Resources Inventory, Volume 1 (1986); Volume 2 (1986)
Sustainable Agriculture and the Environment in the Humid Tropics (1993)
Sustainable Agriculture Research and Education in the Field: A Proceedings (1991)
Toward Sustainability: A Plan for Collaborative Research on Agriculture and Natural Resource Management (1991)
Understanding Agriculture: New Directions for Education (1988)
Use of Drugs in Food Animals: Benefits and Risks, The (1999)
Water Transfers in the West: Efficiency, Equity, and the Environment (1992)
Wood in Our Future: The Role of Life Cycle Analysis (1997)

Nutrient Requirements of Domestic Animals Series and Related Titles

Building a North American Feed Information System (1995)
Metabolic Modifiers: Effects on the Nutrient Requirements of Food-Producing Animals (1994)
Nutrient Requirements of Beef Cattle, Seventh Revised Edition, Update (2000)
Nutrient Requirements of Cats, Revised Edition (1986)
Nutrient Requirements of Dairy Cattle, Seventh Revised Edition (2001)
Nutrient Requirements of Dogs, Revised Edition (1985)
Nutrient Requirements of Fish (1993)
Nutrient Requirements of Horses, Fifth Revised Edition (1989)
Nutrient Requirements of Laboratory Animals, Fourth Revised Edition (1995)
Nutrient Requirements of Poultry, Ninth Revised Edition (1994)
Nutrient Requirements of Sheep, Sixth Revised Edition (1985)
Nutrient Requirements of Swine, Tenth Revised Edition (1998)
Predicting Feed Intake of Food-Producing Animals (1986)
Role of Chromium in Animal Nutrition (1997)
Scientific Advances in Animal Nutrition: Promise for the New Century (2001)
Vitamin Tolerance of Animals (1987)

Further information, additional titles (prior to 1984), and prices are available from the National Academy Press, 2101 Constitution Avenue, NW, Washington, D.C. 20418, 202–334–3313 (information only). To order any of the titles you see above, visit the National Academy Press bookstore at http://www.nap.edu/bookstore.